# The Power of the Eyes

**Patricia Webbink,** Ph.D., is a psychologist in private practice in Bethesda, Maryland. Twenty years of research on the eyes, including her dissertation at Duke in 1974, have generated the subject matter for this book. The author has taught university courses and lectured on this and other topics both here and abroad. Furthermore, she has contributed chapters on the eyes and imagery in several publications. Dr. Webbink uses the imagery of the eyes in her private psychotherapy practice. Her work on a new approach called centering has recently been featured on the Voice of America, several times on the local news, and on the front page of the *Wall Street Journal.* The author currently teaches at the University of Maryland, and has been invited by the university to present seminars for business people on the "problem employee" and other management issues. In the past she has taught at George Washington, American, and Duke Universities.

# The Power
# of the Eyes

Patricia Webbink, Ph.D.

SPRINGER PUBLISHING COMPANY
New York

Springer Publishing Company, Inc.
536 Broadway
New York, NY 10012

86 87 88 89 90 / 5 4 3 2 1

**Library of Congress Cataloging-in-Publication Data**

Webbink, Patricia.
  The power of the eyes.

  Bibliography: p.
  Includes index.
  1. Eye contact.  2. Interpersonal relations.
3. Therapist and patient.  4. Eye contact—Social
aspects.  I. Title.
BF637.N66W43  1986       153.6       86-1744
ISBN 0-8261-2670-7

Printed in the United States of America

Rivers of color,
Mirrors of pain,
The sun sifts through
     and touches other souls.
Penetrates and merges,
Sparkles with laughter
     and sprinkled with tears.
Opening with wonder,
Closing with shame.
We can share our depths
     or escape with our eyes
          in a millimeter of flight.

Patricia G. Webbink
11/25/85

# Contents

# Acknowledgments

This book represents fifteen years of research in the field and has been influenced by many people, beginning with my professors in graduate school at Duke University. During the time that I was investigating eye contact for both my master's thesis and doctoral dissertation, Robert Carson, Ph.D., my dissertation advisor, spent many hours with me discussing this topic. I am especially grateful to Norbert Enzer, M.D., who gave generously of his time and wisdom, and the late John Thibaut, Ph.D., of the University of North Carolina, who encouraged me to write this book. I then approached Ursula Springer, President of Springer Publishing Company, with the idea. In the course of working on it she has become a close friend and respected colleague, and her continual encouragement and excitement about the subject helped to make the book a reality.

Many others have assisted with the writing of this text, some of whom I may inadvertently fail to mention. I am grateful to all of them. Several years ago Joanne Malone and Janice Eckland worked on the beginning of the book. Diane Fetter, Ph.D., has contributed to discussions on the art symbolism section. Carol Wyrick helped with a number of parts, including the "Evil Eye" and the section on the politics of eye contact. Ralph Exline, Ph.D., has talked with me over the years, and has helped to keep me informed about the most current research. To Allan M. Leventhal, Ph.D., many thanks for his support throughout this project.

By far the most significant contributions have come from Toni White and Jane Noll, who worked diligently and with dedication over the past few years, researching, writing, editing, and helping with almost every aspect of the book.

The final manuscript was the result of the work of many typists: Donna Bachtell, Kristy Campbell, Terri Crites, Rosalind Hall, Marilyn Schoenfelder, and Phyllis Taylor. Readers have included Alison Bennett, Audra Stone, and Allen Lenchek.

I wish to thank all of the people, mentioned and unmentioned, who gave generously of their time to help make this book a reality.

# Foreword

It might be said that this book is the product of two decades of the author's interest in the eyes and their place in human communication. It is, indeed, a compendium of Dr. Webbink's analysis of the work of many investigators and clinicians as well as her own research and clinical experiences. It is a summary of scientific work which should be a significant mile post in the history of the study of nonverbal communication. As such, it will serve as a resource for investigators and clinicians to promote greater understanding and to stimulate further work. But it is also much more.

Dr. Webbink has incorporated into her scientific review the folklore of eyes in a manner that enlivens the narrative yet does not confuse research with legend. The clinical vignettes and real life examples add practical richness. The book is, therefore, much more than a review of the experimental studies of the role of the eyes in human behavior and in interpersonal relationships. It is perhaps best described as a series of essays on human relations that draw on diverse and numerous sources: the literature of the social and behavioral sciences, the arts, politics, the news media, and even comic strips. Each chapter might virtually stand on its own, yet each creates an awareness of the expressiveness of the human eye in particular social contexts, of the capacity of humans to convey their feelings, their intentions, or their positions with their eyes, and of the power of the gaze. Collectively these chapters provide a literate, comprehensive look at one aspect of the human condition which, though a shared experience for most, is often passed over lightly or poorly understood. As Dr. Webbink points out, we often speak of the eyes or the gaze symbolically, attempt to interpret the gaze of others, or to communicate specific messages with our eyes. There does seem to be a language of the eyes which, like the spoken language, may have its own syntax. Although this volume can scarcely be viewed as a text on the grammar of the eyes, it can be seen as a recitation of the style of this form of nonverbal communication.

This book is more than a textbook. It is a compassionate, insightful commentary by a sensitive, scholarly clinician which can enrich the lives of readers, whether scientist or lay person, clinician or client.

I cannot conclude without a brief personal comment about the author, for it was just about twenty years ago that I first met Dr. Webbink. At the time, she was an enthusiastic, creative, and challenging graduate student in psychology who seemed to have almost boundless energy. Although our contact in more recent years has been limited, it is apparent from this book, as well as from her busy clinical practice and her role as a mother, that she has maintained those characteristics.

NORBERT B. ENZER, M.D.

*Professor of Psychiatry*
*College of Human Medicine*
*Michigan State University*

*This book is dedicated to
my son, Andrew, who has magical eyes,
and to the memory of John Thibaut, Ph.D.
whose original idea it was to write this book,
and who gave me inspiration and encouragement.
John was a wonderful human being
and outstanding psychologist.*

# INTRODUCTION:

# *The Eye Opener*

The purpose of this book is to explore the many aspects of a specific type of nonverbal communication called eye contact, the mutual interaction that occurs when two pair of eyes meet. As the evidence shows, eye contact plays a powerful role in interpersonal bonding; and in a world characterized by mechanization, threats of violence, and social alienation, the need for increased interaction between people is apparent. The power of eye contact is very real: Mutual gaze is a major form of communication which promotes intimacy. It facilitates a deeper, more intense knowledge of the other person and, by reflection, of ourselves. Understanding the how and why of eye contact—its use in social interaction, its development from infancy, its function within the power dynamics of society, etc.—can help us to increase the effectiveness of communication with others and to enhance feelings of closeness with those around us.

It is hoped that this book will increase awareness of the power of eye communication. Enriched by such awareness, we will be able to find new ways of making contact with others; clinicians and teachers will know more about their clients and students, and will be able to provide more effective therapy and teaching; individuals communicating across cultural boundaries will recognize a potential source of prejudice and misunderstanding. In this book, social, clinical, physiological, and experimental psychologists should find possibilities for further research pertinent to social interaction, clinical practice, and social change.

The special perspectives and wholistic values of the author make this book on eye contact different from others in the area, and have affected the methodology, structure, and concepts presented here. The total human context surrounding eye behavior is examined with the view that the behavior of people in laboratory settings reveals only some of the important factors of human experience. The capability of psychological experimentation to provide quantifiable facts and reliable information about eye contact is not to be underestimated. However, experimental research is only one source of knowledge (Wilbur, 1983). Culture, phenomenology, and symbolism are explored as well (Chapter VI).

A feminist approach is evident in the language of this book. "S/he" (an abbreviated form of "she or he") is used instead of the commonly accepted generic "he," in order to prevent sexist assumptions which further female invisibility. However, authors quoted in this text who use the generic "he" are quoted as originally published in order to avoid the constant insertion of "(sic)." The reader should not, however, assume that such passages refer only to men, unless otherwise indicated.

The author's political and sociological approach to psychology is most apparent in Chapers II and III, which examine patterns of eye contact within the context of unequal power relationships. The author's goal is to assist social change that creates an environment that fulfills people's needs for self-actualization and mature relationships.

In translating the language of the eyes, centuries of evolving folklore have created a variety of meanings associated with eye movement, color, and shape. The task of translating a nonverbal communication system into verbal expressions is by its very nature a difficult one (Scheflen, 1978). Interpretation of the more subtle gestures of the language can be a formidable task. For example, laboratory studies over the past twenty years have resulted in contradictory and inconclusive findings regarding the accuracy of an observer's ability to perceive eye contact as it occurs between experimental subjects. (See Appendix I, "Measuring Eye Contact," for a review of some of the research problems involved.)

After determining how eye contact fits within the overall system of verbal and nonverbal communication systems, Chapter I, "The Language of the Eyes," delineates the presently known and experimentally verified functions of eye contact as a language of human social interaction.[1] Research psychologists have established that eye contact serves to signal one's own attention and to elicit the attention of others; regulates turn-taking in social interaction; and expresses interpersonal attitudes.

While understanding that there are what psychologists call "normal" patterns of mutual gaze, one must keep in mind that certain factors cause variations between individuals and between different cultural and subcultural groupings. It is also important to be aware that much of the research contributing to the concept of "normal" eye behavior has been conducted with white college students, who comprise the majority of experimental subjects (Braginsky & Braginsky, 1974). Therefore, Chapter I ends by probing cultural differences in eye contact. In a shrinking world where people of many cultures are interacting with one another, differences in body language are as important as differences in verbal and written language. For those who wish to overcome barriers in intercultural communication, knowledge of cultural eye contact patterns is crucial, as is knowledge of other culturally determined nonverbal behaviors. Fehr and Exline (in press) have emphasized the usefulness of cultural comparisons because "(they) frequently provide a much needed contrast to our own interaction patterns, enabling us to demystify what seems natural" (p. 130).

In Chapters II and III, eye contact is examined in relation to the distribution of power in society. The kind of looking which occurs between people reflects not only the level of intimacy between them (see Chapter IV), their personality differences (see Chapter V), and their cultural backgrounds, it also reflects their social roles. Status, power, dominance—factors determined by political ideology and social systems—affect how and when one person will look at another. Chapter II begins with an examination of how charismatic persons in positions of power use their eyes to exert a special influence and manipulate interactions. The chapter closes with a description of the specific eye behaviors associated with the roles of dominance and submission.

The examination of gender and minority differences in a political context, as in Chapter I, helps us to avoid biological determinism as an explanation for the differences in looking behaviors between the sexes and between members of dominant majorities and minorities. As Patterson asserts: "We have to appreciate that it is not sex per se as a variable that causes these differences, but some differences in socialization" (1978, p. 19).

Many movements have been concerned with the effects of body language and other nonverbal signs. For example, dress, music, and dance have been very important to the creation of group identity and a sense of power by groups working for increased freedom. As awareness is changed by the adoption of open-minded beliefs, many have begun to change their nonverbal as well as verbal behaviors in an effort to create egalitarian relationships. As we begin to succeed in creating a world of justice for all, lowered heads and furtive glances will be discarded; equality will produce an atmosphere where we can look each other in the eye. For those interested in breaking down sex, race, class, age, and other barriers in our society, the information in Chapters IV and V on how eye contact supports present inequalities will be of considerable interest.

Psychologists and their students should find Chapter IV, "Eye Contact and Intimacy," a fertile ground for further research. In modern technological and bureaucratic society, people's needs for intimacy are often ignored. Nonverbal communication is a powerful channel for intimacy between people. In particular, eye contact provides simultaneous and reciprocal experiences of mutual connection. It is this aspect of eye contact which has intrigued the author for years. Personal experiences of intense union through eye contact have led the author to feel that eye contact can be extremely significant to human relationships. Eye contact can help us break through barriers to communication, and has been correlated with such aspects of intimacy as honesty, love, and empathy.

Because of its widely variable nature, eye contact often fluctuates with changes in the level of intimacy between people. The research of Argyle and Dean (1965) indicates that as intimacy increases between people, the amount of eye contact will decrease in order to maintain an "equilibrium" level of intimacy. Their research has generated much discussion in the experimental literature on eye contact. Chapter IV explores "equilibrium theory," presents contradictory findings, and offers Patterson's "arousal model of interpersonal intimacy" (Patterson, 1976, 1978; Patterson et al., 1981) as a possibly more encompassing way to look at the relationship between eye contact and intimacy.

The roots of the relationship between eye contact and intimacy can be found in the interplay of eye behavior between the infant and her/his caregiver.[2] For the infant, eye contact is perhaps as important as food itself in that it provides the earliest experiences of mutual, reciprocal union with another being. For the caregiver, mutual gaze with an infant often releases caregiving, affectionate responses. It makes the relationship significantly meaningful for the caregiver, providing a sense that the caregiver is being loved.

Several psychologists have stated that the nature of early eye contact experiences determines the nature of later adult relationships and emotional development. Many researchers have focused on the mother and her behavior as the major determinant of the child's or infant's behavior. Recently, however, some have observed that the infant or child also exerts a major influence on the infant/caregiver relationship; children can actually control interactions through the giving and withdrawing of eye contact. Chapter IV emphasizes an "interactive" approach to the study of eye contact in infant/caregiver relationships.

Although it seems apparent that early experiences with eye contact will influence later eye behavior and openness to intimacy, more research is needed in this area before any conclusive statements can be made. There are huge gaps in our knowledge of the development of eye contact throughout a lifetime. Chapter IV will hopefully stimulate some researchers to fill these gaps.

Chapter V, "Clinical Aspects of Eye Contact," provides information about psychological problems. This discussion is unique in specifically dealing with the clinical issues of eye contact, rather than simply including the gaze patterns of psychiatric populations in a survey of individual differences. Eye contact has usually been investigated by social or experimental psychologists; the therapeutic, diagnostic, and teaching aspects of eye contact have previously received minimal attention. In Chapter V, eye contact is explored within a clinical context and is amply illustrated with material from case histories.

Within a given culture, it is apparent that people with severe psychological problems often deviate from normal gaze patterns. Such symptoms as closed eyes, gaze aversion, elevated blinking rates, decreased eye contact, fixed stares, and/or unusually prolonged gazes can be used as diagnostic tools. Autism, schizophrenia, depression, multiple personalities, anxiety, tension, and stress are often manifested by characteristic gaze patterns. Not only can a therapist use eye behavior as an aid to diagnosis, s/he can also monitor a client's progress by assessing changes in eye contact.

Besides its usefulness as a diagnostic tool, an indicator of progress, and a means of observing changes throughout psychotherapy, eye contact between client and therapist can also have healing properties. Although some therapists may not be aware of how their nonverbal behavior is affecting their clients, evidence shows that eye contact is a primary means of expressing empathy, which is the attitude most often linked to therapeutic effectiveness. Eye contact can serve to intensify almost any therapeutic experience. Chapter V discusses how it can be used in a variety of therapeutic techniques, including psychodrama and group therapy. Practical uses of eye contact in psychotherapy are suggested, and it is hoped that these suggestions will be of use to clinicians.

Chapter VI, "Eye Symbolism," provides a unique examination of the sociocultural heritage of the eyes throughout history. [There is only one piece of extensive work on the symbolism of the eye, a lengthy and comprehensive book in German, *Urmotiv Auge* (1975), by Otto Koenig.] The evidence of mythology, religion, visual and dramatic arts, language, literature, folklore, and custom reviewed in Chapter VI makes it apparent that the eyes have provided a rich poetry of images and meaning in human consciousness through the ages. As a major element of symbolic systems, the eye image has been universal in its illumination of certain aspects of human experience.

Advertisers take advantage of the emotional expressiveness of eyes, often using sets of eyes to show the emotional response that the advertisement attempts to elicit, or to express how the potential buyer will feel about a product of its price. Advertisements for movies, books, and plays use eyes to convey quickly and succinctly the emotions that will be elicited by the story. The posters, sheet music, and album cover from the recent play "Cats" are illustrated with a black background containing only a pair of cat's eyes. Through the use of eye illustrations the mood of a story can be communicated before it is even read. Popular belief invests the eyes and eye contact with a kind of mystical power by regarding them as mirrors of the soul. The eyes are believed to be a special source of truth about a person; they are said to have a language of their own which "transcends speech" (Emerson, in Bradley et al., 1969), and there is worldwide belief that the eyes do indeed express a great variety of emotions and character traits. Writers, actors, visual artists, and advertisers have used eye expression and eye images throughout history as a primary mode of communication and representation because of our belief in the eyes as windows of the soul. Like other powerful symbols, however, eyes have also been associated with certain destructive aspects of human culture. For example, visually-impaired persons have often been stigmatized. Beliefs about the power of the eyes have also functioned as the means for societies to purge themselves of "undesirable" or "eccentric" persons by labeling them as possessors of the "Evil Eye."

Even a book as broad in scope as this cannot include the innumerable insights and observations made about eye contact nor all the research which has slowly accumulated in the field. The interested reader is urged to continue her/his study and to examine the work of various people who have made substantial contributions to our understanding of mutual gaze. For example, Dr. Ralph Exline at the University of Delaware, both alone and in collaboration with many colleagues, has been working for over twenty years on quantifying a variety of individual differences in looking behavior (1962, 1963, 1971). He and his colleague, B. J. Fehr (who is now at Tufts University), have written comprehensive reviews of the experimental literature on social–visual interaction which is soon to be published. Some of the very early significant work in this field was done in Europe by Gerhard Nielsen (1962) in Denmark and Adam Kendon (1967, 1968) and Michael Argyle (1968, 1969, 1970, 1973, 1975), both of whom were in England. Argyle and Dean (1965) have formulated a pioneering analysis of nonverbal behavior, the "equilibrium theory," and Argyle and Cook (1976) have provided the first comprehensive review of the psychological literature on eye contact in a book called *Gaze and Mutual Gaze*. Unfortunately, the only extensive work on eye symbolism, Otto Koenig's *Urmotiv Auge* (1975), is not available in English.

Three other individuals deserve mention as well, for although less extensive, their work is original, creative, and innovative. Nancy Henley's *Body Politics* (1977) includes a chapter on eye contact and, along with numerous articles, examines how nonverbal behavior expresses personal and political power. John Heron's phenomenological and philosophical treatment (1970) of eye contact stands out for its in-depth, detailed inquiry into the nature of the gaze. In "The Phenomenology of Social Encounter: The Gaze," Heron makes the distinction between merely looking at a person's eyes and looking into them. Silvan Tomkins, in *Affect, Imagery and Consciousness* (1963), also explores the complex, qualitative aspects of mutual gaze in social interaction. Each of these unique perspectives has provided us with a major contribution to the field.

Although scientists are perhaps best equipped to quantify or measure eye contact, certainly this book shows that they are not the only ones concerned with its use in daily life. Because eye contact is integral and essential to most social relationships, it should be of concern to anyone interested in the human condition or, indeed, simply in the condition of one's own life. While we may not all see eye-to-eye about its nature or functions or measurement, surely we can agree on the importance of mutual and reciprocal gaze in revealing who we are to one another.

## NOTES

[1]For those who wish to refresh their understanding of the biological aspects of visual communication, Appendix I reviews the human visual system and discusses the evolution of visual systems in animals.

[2]The word "caregiver" is carefully chosen here to indicate that not only mothers are involved in the creation of early relationships with infants. Fathers, other relatives, friends, and childcare workers are also major influences in children's lives. Unfortunately, most research has focused only on the relationship between mothers and their children.

# CHAPTER I

# *The Language of the Eyes*

This chapter explains, with the detail that the lens of psychological experimentation reveals, how eye contact functions in human communication. The interplay of eye contact between persons is part of the nonverbal language that human beings use to coordinate activities and convey meaning to one another. When it comes to our understanding of eye language, most of us are like children just beginning to talk. We imitate or ridicule the language of the eyes of those around us without knowing the rules of "grammar" or the meanings of the "words" we are saying with our eyes. In the past twenty years, however, research in experimental psychology and other social sciences has produced data which we can use to increase our command of eye language.

Like other forms of communication, eye language is created and used in specific cultural contexts which make its rules or usage and its meaning unique to those cultures. This chapter closes with a discussion of the universal and culture-specific aspects of eye language, offering help to those involved in intercultural communication.

## NONVERBAL COMMUNICATION

As human beings, we are social creatures, and have often organized our behavior in relation to one another. In order to organize social activities among ourselves, we have had to develop systems of communication. For communication to occur between two or more persons, the information, ideas, and/or emotions to be transmitted were first encoded into "signs" which have a similar meaning for those involved in the interaction (Fehr & Exline, in press). Through the evolutionary development of our species, the historical development

8

of our culture, and the personal development of every individual, a shared code for organized communication has been developed (Fehr & Exline, in press).

The signs that we emit and perceive are expressed in two modalities, verbal and nonverbal. *How* we choose to express something is related to *what* we wish to communicate. The verbal mode is most often used to convey cognitive information, while the nonverbal mode is used to express emotion (Trout & Rosenfeld, 1980). Thus, "nonverbal codifications represent an intimate language," while the verbal ones represent a more distant language (Ruesch, 1972, p. 730).

In mainstream United States culture, social expectations often require the inhibition of verbal expression of feelings (Trout & Rosenfeld, 1980). Nonverbal signs of emotion are therefore assumed to be more indicative of a person's true affective state and less subject to deception than verbal expressions. It is not uncommon to have an experience in which someone tells us s/he is "feeling fine," while her or his unconvincing voice tone, lack of direct gaze, and hunched posture tell us a different story. In such an encounter, we can choose to respond to the verbal content or to the body language—but, in any case, we are apt to conclude that the person's "true" state is not "fine" at all. As Fast noted in his popular book, *The Body Language of Sex, Power, and Aggression* (1977), "body language is . . . an unconscious language, and because of that it is a very honest language" (p. 10). Fast's assertion about the unconscious nature of nonverbal behavior should be qualified, however. We all have some degree of awareness about the meaning of our nonverbal expressions. For example, anyone who attempts to lie or conceal information probably will be conscious of trying to "act" as if s/he is telling the whole truth.

Nonverbal communication comprises 65% of all communication (Kundu, 1976). Ruesch (1972) asserted that "the basis for human relations is established in the nonverbal mode" (p. 638). Yet, psychology has directed its attention to nonverbal behavior only relatively recently (Knapp, Wiemann, & Daly, 1978). In the 1950s behavioral scientists finally began to recognize the importance of nonverbal behavior. They started to translate body language systematically and to study the cultural rules regulating its use in human communication. Before these studies, nonverbal communication was largely an "elaborate and secret code . . . written nowhere, known by none, and understood by all" (Sapir, 1927, p. 137). A number of popular books on the topic appeared in the 1970s and caught the attention of the general public. Best sellers like Fast's *Body Language* (1970) created a new mass consciousness of nonverbal behavior. In the Foreword to a later book, Fast (1977) noted, "wherever I talked . . . hundreds of people—students, children, husbands, wives—pressed me for answers to very personal questions. . . . They saw in body language, a means of getting a little closer to each other, of gaining some meaningful insights, of communicating on a deeper, more honest level" (p. 11).

This popular interest in nonverbal communication reflects the expanding amount of solid research literature which is now culminating in major syntheses of existing data. Results are becoming available from long-range studies, and theoretical issues are being clarified (Harrison, Cohen, Crouch, Genova, & Steinberg, 1972; Harper, Wiens, & Matarazzo, 1978; Fehr & Exline, in press). Yet, there is no real consensus as to the exact definition of "nonverbal communication," the extent of the field, or the best research approaches (Harper et al., 1978).

Technically, the term "nonverbal communication" refers to all human responses which are not manifested as words (Knapp, 1972). Thus, we could apply the term "to a broad range of phenomena: everything from facial expression and gesture to fashion and status symbol, from dance and drama to music and mime, from flow of affect to flow of traffic, from the territoriality of animals to the protocol of diplomats. . . ." (Harrison, 1973, p. 93). Psychologists have attempted to narrow their field of study in nonverbal behavior to include those aspects which appear crucial to the structure, occurrence, and regulation of interpersonal interaction (Harper et al., 1978). Among the important nonverbal behaviors which have been identified and studied are: body movement (kinesic behavior); proxemics (the use of space)[1]; paralanguage (e.g., voice tone, verbal timing, and accent); facial expression; and eye behavior.

Certain dynamics or fluctuating behaviors (utterances, movements, looks, etc.) serve to pattern the flow of an interaction in time (Welford, 1958). Dynamic features coordinate the alternation of active message-sending by participants. It has been suggested that this patterning function is crucial to the outcome of social interactions (Chapple, 1940).

Greeting and leave-taking rituals illustrate how nonverbal cues can function to pattern interactions. The filmed studies of Kendon and Ferber (1973) showed the greeting ritual to contain a series of phases utilizing gaze signals and other nonverbal cues. First, before any obvious sign of greeting is given, the two interactants sight each other and begin to synchronize their movements. Then, a head toss, nod, or wave is offered as salutation from a distance. As the interactants approach, eye contact occurs, they smile, and a hand may be offered. When a person wishes to end an interaction and leave, "buffing" occurs; i.e., short sounds like "ah" and "er" are uttered to change the discussion (Knapp, Hart, Frederich, & Shulman, 1973). Eye contact is broken, bodies position themselves to the left, and there is forward lean and head nodding (Knapp, Hart, Frederich, & Schulman, 1973). Kendon and Ferber (1973) asserted that only when departure moves are matched by the other interactant can leave-taking be accomplished. They also observed that just before the verbal "goodbye" there is a head toss combined with mutual gaze and a wave.

Nonverbal behavior serves as an expressive companion to spoken communication. Ekman and Friesen (1969) listed four such functions of nonverbal behavior, described here by Harper et al. (1978):

> The most obvious function is perhaps *repetition*, as when a baseball umpire may signal "out" with his thumb while simultaneously verbalizing. Verbal and nonverbal behaviors may also *contradict*, as in the case of verbal praise given in a sarcastic tone of voice. Nonverbal behaviours may also *complement* verbal behavior: praise can be given with a smile or pleasant tone of voice. In addition, they can even *accent* spoken words. One can lean forward or touch the other interactant in both verbal and nonverbal display of affection. (p. 7).

In 1964, Birdwhistell reported his observations of the close relation between speech and body movement. Specifically, nods or sweeps of the head, eye blinks, brow movements, and hand or finger movements occur in association with points of linguistic emphasis (Argyle & Kendon, 1967). Argyle and Kendon (1967) suggested that such use of nonverbal cues may reduce ambiguity for the listener and may keep the listener's attention by breaking the conversation into small chunks.

All the dimensions of nonverbal expression, including touching, movement, body distance, paralanguage, facial expression, and eye behavior, are interrelated and form a whole picture. The overall configuration of signs emitted by interactants must be considered as a whole if we are to understand the complete meaning of an episode of communication. Different nonverbal cues provide different kinds of information and the more nonverbal cues we have from others the more personalized and spontaneous our interaction can be (Rutter & Stephenson, 1979). It can be quite frustrating trying to have an emotional conversation over the telephone when vocal paralinguistic expression is the only nonverbal mode of communication available to us. For impersonal, businesslike exchanges, however, using the telephone may feel quite adequate.

Although the interrelation of all forms of nonverbal behavior cannot be overlooked (Schwartz, 1982), some have observed that head cues are more important than body cues in the receiver's perception of the sender's attitudes (Mehrabian, 1968b). Ekman and Friesen

(1967) stated that body cues are more ambiguous than head cues. Buck, Savin, Miller, and Caul (1972) found that emotions can be accurately communicated through facial expressions alone. The face is probably the most compelling channel of nonverbal communication and is so important to most of us that we equate our identity with the appearance of our faces (Weitz, 1979). Assumptions about a person's character are often made from facial appearance, and injuries to the face are usually seen as a particularly awful type of disability (Montagu, 1971; MacGregor, 1974), thus affecting a person's self-concept and potential for social survival (Weitz, 1979). During social interactions we pay almost exclusive attention to the face. Albert Mehrabian (1968b) determined that facial expression plays a greater role in determining the total impact of a communication than verbal and other vocal behaviors combined. Other studies have shown that most people tend to concentrate on the facial area when looking at a picture of a person (McLuhan, 1964).

## THE FUNCTIONS OF EYE CONTACT
## IN HUMAN COMMUNICATION

Our selective attention to facial cues is accompanied by a focus on the eyes. Argyle (1968) speculated that there is an "instinctive attraction of the eyes" (p. 109) which draws us to them when we are looking at someone's face. Fast (1977) asserted, "The eyes are the parts of the body used most often in nonverbal communication" (p. 116), and Simmel (1924), who is often quoted, suggested that the eyes play a unique social function, going so far as to state that "the union and interaction of individuals is based upon mutual glances" (p. 358). Our tendency to focus on the eyes has been documented by the research of Janik and associates (1978). Using an eye tracking camera, they found that while visually inspecting facial photographs, male subjects spend more than 40% of their looking time focused upon the eye region. The remaining portions of the face were looked at for relatively short periods. (The mouth, the next most popular area, was looked at more than 10% of the time.)

The eyes do indeed have a particular saliency in the environment. We are far more likely to notice a direct gaze before subtle postural changes (Ellsworth, 1975). Despite the quietness and subtle movement of the direct gaze, we expect people to respond to it, and are frustrated when they do not, as when a waiter avoids our look in a restaurant (Ellsworth, 1975).

Within the total communication system embodied in human social interaction, eye behavior constitutes a communication subsystem. Like nonverbal behavior in general, eye contact is a language which expresses the emotional qualities of our relationships with one another. It also serves to regulate our interactions and to qualify our verbal responses. On the receiving side of eye communication, gaze can inform us of the current orientation of others, their underlying predispositions (see Chapter VI), transitory states, and "potentially their next move, providing us with the capability of adjusting to it" (Fehr & Exline, in press, p. 20). Eye contact is a language of relationships, helping to define who interacts, how they interact, and what they communicate.

## The Eyes Catch You

If you want to get someone's attention so that you can interact with him or her, catch his or her eyes with your own. Hitchhikers use this method of obtaining a ride. Waiters and waitresses use eye contact as an acknowledgment of their readiness to serve or its avoidance as a communication that they are too busy to respond. In addition, one's emotional reactions to other passengers in a subway car will vary greatly according to whether they are looking into our eyes when we glance at them or not. If they should look into our eyes and maintain that gaze, we would probably become anxious and search for some clue to the reason for their behavior.

Seeing two eyes directed toward our own is a powerful trigger for response. In infants, the image of two eyes is the minimum visual stimulus necessary to elicit a smiling response (Spitz & Wolf, 1946; Ambrose, 1961). Studies have shown that being gazed at also causes increases in people's EEG arousal (Gale, Lucas, Nissin, & Harpham, 1972; Gale, Spratt, Chapman, & Smallbone, 1975), GSR (Nichols & Champness, 1971), and heart rates (Kleinke & Pohlen, 1971).[2]

Cross-cultural beliefs in the power of the eyes (see Chapter VI) and the responsiveness of animals to eyes and eye-like configurations (see Chapter II) support speculation that there is probably an innate basis for our being drawn to others' eyes (Argyle, 1968). Eibl-Eibesfeldt (1972) identified a universal pattern of greeting behavior which is probably an innate response (Harper et al., 1978). This greeting pattern, occurring in about one-third of a second, consists of directing gaze toward another person, smiling, lifting the eyebrows, and quickly nodding the head. Acting as a "releaser," this brief series of acts elicits the same behavior from others (Harper et al., 1978). As Kendon and Ferber (1973) noted, greetings serve "an important function in the management of relations between people" (p. 592).

Unlike other sensory receptors, it is necessary to direct the eyes toward the object of perception; the physical aspects of visual behavior therefore contribute to attributions of attention and orientation (Fehr & Exline, in press). When two sets of eyes meet, there is mutual recognition that communication may begin (Hutt & Ounsted, 1966); the signalling channel is opened. This opening of the channel of communication through mutual gaze is twofold. When the gaze of another is upon us we become aware that this person is attending to us. At the same time, the gaze of the other elicits our attention toward her/him.

Since direct gaze signals involvement between persons and a mutual recognition of humanness, gaze avoidance denies another's presence and makes her or him feel like a "nonperson." Celebrities can avoid offending their admiring fans by using dark glasses to ignore strangers who might otherwise catch their eye and try to involve them in unwanted conversation (Fast, 1977).

Gaze aversion is also associated with the termination of an encounter. Knapp, Hart, Frederich, and Schulman (1973) recorded the occurrences of verbal and nonverbal behaviors during the 45 seconds before an interactant arose from his chair to leave. (All participants were male.) Of the nonverbal behaviors observed, the breaking of mutual gaze was observed to occur the most frequently.

Goffman (1963) suggested that "civil inattention" is the basic rule governing gaze behavior among strangers. The "rule" requires people initially to offer visual attention, thus acknowledging the other's existence, and then to withdraw it to indicate that the person is of no special concern. As Cary (1978) pointed out, "By giving visual attention in excess of a first look a person can indicate that an encounter is desired without actually speaking . . ." (p. 269). Cary (1978) studied undergraduate students' behavior in a laboratory waiting room. Videotapes of interactions showed that when a person entered the room while another waited within, the person waiting almost always gave initial visual attention when the door was opened. If such attention was not given, conversation did not occur; however, visual attention did not necessarily guarantee conversation. As the entering person moved away from the door, another mutual look greatly increased the chance of an ongoing conversation and made the possibility of no talk occurring highly unlikely (Cary, 1978).[3]

Gaze functions as a signal for social action in groups as well as in dyads. Callan, Chance, and Pitcairn (1973) reported that gaze played a central part among members of a therapy group in signalling various levels of openness to interaction. The "alert" person gazed at others and was open to interaction; the "huddled" person avoided visual contact and demonstrated withdrawal from the group. The "closed" person averted gaze and did not volunteer any conversation unless asked, while the "away" person stared in an unfocused fashion into space. Weisbrod (1967) concluded that to look at someone while s/he is speaking in a group is a sign that one wishes to be included by the speaker in the discussion. Receiving looks back from the speaker is taken to mean that one is being included by the speaker.

Although we can generally interpret the gaze of another toward ourselves as a signal of their attention, it is possible for someone to look in one direction while actually attending to something or someone else through hearing or peripheral vision. A person can also be lost in thought, gazing at but not really seeing the objects and/or persons in their visual field. Thus, Callan, Chance, and Pitcairn (1973) differentiated the official display of attention ("advertence") from the actual reception of stimulus input (real attention). We could be "fooled" by the visual advertence displays of others, but there may be other cues in the overall behavior of a person which help us correct possible misinterpretations (Fehr & Exline, in press).

## Eye Contact Regulates Interaction

Along with gestures, eye contact functions as a primary regulator of human social interaction (Ekman & Friesen, 1969). "Regulation" means the manner in which we are able to communicate smoothly, with each person taking turns and without impinging upon each other's conversational space. Eye contact is an important nonverbal mechanism by which interactants conduct the flowing exchange of speaker and listener roles in a conversation. It has been suggested that this patterning function of nonverbal behavior is crucial to the outcome of social interactions (Chapple, 1940).

In order for us to regulate our behavior smoothly in a conversation, we must continuously adjust our behavior to others' as they adjust their behavior to ours (LaFrance & Mayo, 1978). People rely on eye signals to help regulate their interactions and the lack of such cues contributes to a breakdown of the flow. For example, Argyle, Lalljee, and Cook (1968) found that synchronization was worse if eyes were concealed by dark glasses. Eye contact helps to regulate the smooth maintenance and exchange of roles in a conversation because it provides a crucial link in the communication process: feedback. In a social interaction, an individual performs a sequence of behaviors which are directed towards a goal. In order to bring about desired consequences, a person must emit output which relates to the particular input available.

The actions one takes (e.g., utterances) create an impact on the listener. When an individual perceives the impact of her/his action s/he receives "feedback," which leads to changes in her/his cognitions and thus influences the type of action s/he will take next. Argyle and Kendon (1967) described how this "sensorimotor skill model of behavior" plays itself out in a conversation:

> (A person) may be discussing current affairs with an acquaintance, and be concerned perhaps merely to sustain a pleasant flow of talk. He might be on the watch then for signs of emotional disturbance in his acquaintance, which might signal that he had said something that might provoke an argument. At another level, he must be on the lookout for signals that his acquaintance is ready for him to talk or for him to listen. He must make sure his tone of voice and choice of words, his gestures, and the level of involvement in what he is saying, are appropriate for the kind of occasion of the encounter. (p. 56)

Kendon (1967) found that the eye behavior of any two interactants tends to fall into a mutually agreed-upon pattern.[4] Partners in an interaction are, therefore, involved in interrelating their performances, coming to what Goffman (1959) has called the "working consensus" of an encounter. Eye contact provides a very immediate and efficient medium for giving and receiving cues helpful to this coordination of actions. In one glance, an interactant can both gather and provide feedback.

In order for communication to be experienced as "well-functioning" and thus "gratifying," individuals must feel that their messages are being received, and that the receiver's subsequent message or reply is connected to the original message (Ruesch, 1972). In short, in order to have satisfying social interactions, humans require feedback. Ruesch (1972) goes so far as to say that not only do we need feedback, we need it in the same mode as the one in which our messages are transmitted. In other words, when we try to communicate something to another by talking and intermittent gazing, we desire verbal and visual responses in order to feel that we have truly "communicated."

Agreed-upon customs or social norms for conversational behavior also contribute to our ability to interact harmoniously. Researchers (e.g., Kendon, 1967) have found that in Britain and the United States certain sequences of eye behavior are *expected* in an interaction; that is, there is a pattern of eye contact which is considered "normal." In general, we are not highly conscious of the regulatory function of eye contact in our daily interactions. However, if someone's pattern deviates significantly from the norm for one's group, we are likely to notice and will make judgments about her/him as a result (Ellsworth & Ludwig, 1972). Argyle (1973) has compiled data on "normal" looking patterns; the following description is based on laboratory studies.[5]

While engaging in conversation, people periodically look each other in the eyes. Argyle (1973) has observed in his laboratory that the percentage of time each conversant looks into the region of the other's eyes can range from 0 to 100%. However, the average is between 30 and 60% of the time. Glances during conversation vary in duration from approximately 1 to 7 seconds (this is quite long when compared with the fixations of visual scanning which are commonly 0.25 to 0.35 seconds). Although other parts of the face may be glanced at, the eyes are the central focus. Otherwise, people tend to look away from the other's face altogether. When one conversationalist looks at the other, the other may look back, producing eye contact. In a typical interaction, eye contact will occur 10 to 30% of the time, for periods of about 1 second (Argyle, 1973).

Gaze direction and conversational patterns are closely related. While listening, people look more often, their glances are longer, and their glances away are shorter (Argyle, 1973; Kendon, 1967). People look the least while speaking (Allen & Guy, 1977; Cherulnik, Neely, Flanagan, & Zachau, 1979; Dabbs, Evans, Hopper, & Purvis, 1980). Just as persons are about to speak, they look away from each other. At the ends of phrases and sentences they look up briefly; and when they finish speaking, they give the listener a more prolonged gaze (Argyle, 1973). At this point the other person will look away and begin to speak. In groups, the person last looked at by the speaker is the one most likely to gain the floor (Weisbrod, 1967).

Kendon (1967) analyzed seven two-person conversations in order to discover some functions of gaze direction in social interaction. He found that if a person did not look up at the end of an utterance (as is expected when one is finished speaking), the listener failed to respond or her/his response was considerably delayed. An important finding by Kendon regarding looking behavior was the very strong similarity between the visual behavior of any two interactants. There were high correlations between the rate of gaze shifts and the length of gaze. To Kendon it seemed as if each pair of interactants came to a gaze consensus, an unconscious agreement about the length of time each would look at the other. He observed that the average length of time each interactant looked at the other differed between pairs.

Exceptions to "normal" or "average" patterns of eye contact have been noted in the experimental literature. Contrary to Kendon's finding that subjects averted gaze at the beginning of utterances and looked back at the end, Levine and Sutton-Smith (1973) found that looking occurred at the beginning of utterances as well as at the end. They also found that children gazed neither at the end nor at the beginning of utterances. Rutter (1978) found that, in adult two-person conversations, whether the speaker looked at the end of the utterance was not critical in signaling floor changes. A look was frequently not given before the floor changed or it was missed by the listener because s/he was looking somewhere else. And while the ratio of looking-while-listening to looking-while-talking generally is nearly 2 to 1, a finding that has been accepted as "reliable," at least for pairs of white interactants (Argyle & Ingham, 1972), Weisbrod (1965) had observed that in a group context the reverse relationship is true. Group members looked 70% of the time while speaking and just 47% of the time while listening.

Patterns of eye contact in interactions vary according to many variables. What has been described here indicates a general norm of eye contact behavior for specific laboratory settings, and as such does not give us much information about situational or cultural differences (see below). Kendon (1978) suggested that social composition of conversation dyads and the nature of conversational structure may be the cause of variability in the role of gaze in conversational turn-talking. The degree of difficulty of a conversation's process or the material under discussion may make the proportion of looking while listening lower than the proportion of looking while speaking (Fehr & Exline, in press). Further research into the verbal and nonverbal aspects of conversational organization is needed and, as Fehr and Exline (in press) observed, "it is not fruitful to test, experimentally, the role of cues isolated from the context in which they occur" (pp. 65–66).

## The Emotional Language of the Eyes

We can discover the attitudes and emotions of others by reading the language of their eyes. Credited with the ability to express a limitless range of emotion and states of being, the "eyes can weep, laugh, threaten, insinuate, deplore, tease, deny, affirm, confirm, dance, or sparkle" (*Voices*, 1970, p. 105). They can be "alive, dead, vivacious, dull, shifty, evasive, direct, strong, steely, weak, seductive, sharp, innocent, or shrewd" (p. 105). Many people believe that when words fail, the eyes can tell. A woman under hypnosis recalled her grandmother who had died many years before. "She had a stroke and could not speak to me, except through her eyes. She was telling me that she loved me." The granddaughter began to cry as she remembered "those beautiful eyes that reached out to me."

In folklore, certain physical properties of the eyes have been traditionally associated with emotions and character traits, and some social scientists have attempted to discover the eyes' physical correlates of emotional meaning. Such fixed characteristics of the eyes as size, angle, position, shape, and color[6] have been said to influence emotional expression. Unfortunately, the interpretations attributed to features of the eyes are in many cases arbitrary and inconsistent. For example, wide-open eyes indicate puzzlement according to Jobes (1961); but to Hart (1949) they express "frankness or naivete, wonder, or terror, depending upon the accompanying facial contractions" (p. 2). Hovorka and Kronfeld (1908–1909) suggested that prominent eyes express frivolity and good nature, but in the Middle Ages, bulging eyes signified evil and depravity of all kinds, including drunkenness, lechery, and gluttony. In some countries, brows which meet are fashionable, and in others they may be considered a sign that the person is a werewolf or vampire (Jobes, 1961). In Elizabethan England, blue eyes were thought to be a sign of debauchery (de Vries, 1974), while some few others consider green eyes an indication of jealousy (the green-eyed monster, it is called). Thus, we see that ascribing physical correlates to emotional expression is an easy task; measuring them accurately is another matter entirely (Jobes, 1961; Meynell, 1947; Stratton, 1934).

If the more fixed physical aspects of the eyes have been considered their expressive aspects, the complex and varied movements of surrounding facial muscles, the eyelids, and other changeable aspects of the eye certainly have contributed even more to their reputation for emotional expression. Reimer (1955) and Hart (1949) noted that, although the eyes are more set in place than most other parts of the face, they are particularly adapted for emotional activity because of their capacity for intricate expression. Many variations are possible with the movement of the eyeball, secretion of the lacrimal glands, and changes in the contours and surfaces of musculature surrounding the eyes. Ortega y Gasset (1957) suggested that "the eye, with the superciliary socket, the restless lids, the white of the sclera, and those marvelous actors the iris and the pupil, is equivalent to a whole theater with a stage and its actors" (p. 116).

Although a vast array of meanings have been attributed to eye language over the ages, recent studies have shown that people do share many common beliefs about the significance of certain eye behaviors.

There is also a large amount of agreement between how receivers interpret gaze and how senders actually feel, or what they intend to communicate (although accuracy of perception does vary).

The eyes play a large part in the facial expression of emotion (Argyle & Cook, 1976). Nummenmaa (1964) found that certain complex emotions (i.e., pleasure, surprise, and anger) can be perceived only through reading the eyes. These emotions were recognized by raters who observed photographs of the eyes. The main cue appeared to be the way the eyes were opened. After observing themselves and others, Thompkins and McCarter (1964) concluded that excitement and interest are conveyed by eyes which look at and follow the observed with eyebrows down. They also observed the following emotions to be expressed by the eyes and surrounding muscles: Joy is conveyed by smiling eyes with circular wrinkles; distress is conveyed by tears and arched eyebrows; fear, by eyes which are frozen open; shame, by eyes looking down; and anger, by narrowed eyes.

Experimental research has verified some of these observations (Argyle & Cook, 1976). The research finding most often corroborated is that sadness is expressed by a reduction in gaze and looking downward (e.g., Exline, Gottheil, Paredes, & Winklemeier, 1968; Fromme & Schmidt, 1972). When Natale (1977) induced females to feel depressed, they gazed at a female confederate less frequently and for shorter durations than those who had been induced to feel elated.

Simply looking away from someone can serve either positive or negative expressive functions. Averting one's gaze can also act as a potent "cut-off," decreasing emotional stimulation in certain interpersonal situations (Harper et al., 1978). Gaze aversion also helps conceal emotional states. The presence of listener-directed gaze in an interaction can be used to intensify the expression of emotions.

Indeed, the authors of the following two studies argued that gaze is related more to intensity of expression than it is to the specific emotion being expressed. Kimble and Olszewski (1980) and Kimble, Forte, and Yoshikawa (1981) instructed participants (all female) to enact positivity/liking or negativity/anger at a camera and at a male confederate. They were further instructed to portray these feelings either with great intensity or with ambivalence. The results showed that there was a larger percentage of other-directed gaze when the message was intense, as compared to when it was ambivalent.

In order to understand the meanings of any visual act, we must take into account a variety of contextual factors. Many subtle aspects, like accompanying facial expression and pupil size, will influence one's perception of a look. (The pupils dilate when people have a strong interest in or feeling about whom or what they are looking at.)[7]

Nonverbal signals are often "indefinite and inexplicit" because they are a language of our "delicate, elusive, and often complicated" private feelings (Schneider, Hastorf, & Ellsworth, 1979, p. 130). Knapp and associates (1978) cautioned us that we must look at "where a particular behavior occurs in the stream of an interaction in order to attribute meaning more precisely. . . . Like words, some nonverbal cues are abstract and embrace many meanings; some are very concrete and have a narrower range of meaning. . . . Any attempt to separate information from context severely affects its utility and meaning" (pp. 275–276).

Obvious cues by which to judge the meaning of a look are (1) social context (e.g., being in a crowded elevator versus being with a friend in your living room) and (2) the type of interaction in which we are engaged (e.g., combat vs. courtship). A list of many of the other external conditions which may affect the meaning of interpretation given to eye contact is presented in Table 1, which is taken largely from the work of Ekman and Friesen (1969).

Depending on its context, a direct gaze can have unrelated or even opposite meanings (Schneider, Hastorf, & Ellsworth, 1979). For example, empirical research supports two rival hypotheses: "That on the one hand to engage in eye contact with someone is to seek to affiliate with him, and on the other hand it is to challenge him" (Kendon, 1967, p. 59). In either case, it is the situational variables which determine how a gaze will be received; the context of a gaze cues its intent.

To those of us who may have spent time with a couple who have recently fallen in love, it is no surprise to learn that in general, people look more at those they like (Exline & Winters, 1965a). Couples who are in love with each other spend more time in mutual gaze and a higher proportion of their looking is mutual (Rubin, 1970). If a person of the opposite sex combines a friendly expression with eye contact, this often conveys a sexual attraction, especially if eye contact is of long duration (Argyle, 1973). Referred to in the psychological literature as an "immediacy cue,"[8] eye contact often communicates positive, affiliative attitudes between persons. This places eye contact in the same category as such nonverbal behaviors as forward lean, which indicates liking (Mehrabian, 1969b) and openness of arms, which indicates warmth (Machotka, 1965).

It has been found in several studies that those with more positive attitudes toward another had more eye contact with that person (Exline, 1963; Exline, Gray, & Schuette, 1965). When an experimenter interviewed two subjects in a study by Mehrabian (1966), the one who received the most looks from the experimenter felt that the

**TABLE 1-1  Situational Factors Affecting Looking Behavior and Its Interpretation**

| | |
|---|---|
| Settings | |
|   Home, office, interview room | |
| Characteristics of room | |
|   Size, color, temperature, | |
|     furniture, props | |
| Characteristics of interactants | |
|   Sex | Personality/individual differences |
|   Race | Cultural background |
|   Status | |
| Number of participants | |
| Emotional tone of interaction | |
| Topic of conversation | |
| Body cues | |
|   Posture | |
|     Relaxed | Forward |
|     Tense (increased muscle | Back |
|       contraction) | |
|   Position | |
|     Standing | Closed |
|     Sitting | Axial position of shoulders, head, |
|     Supine |   legs, torso, arms |
|     Open | |

experimenter liked her/him. In another study, interviewees gave the most unfavorable reactions to nongazing interviewers, rating them as least attractive, giving them the shortest answers, and sitting farthest from them during a debriefing session (Kleinke, Staneski, & Berger, 1975). Apparently, when we like someone, we are motivated to look into her/his eyes more often and we often perceive the gaze of others as an indication of their liking us. When people avert their eyes from ours we not only feel less liked; in return, we like them less.

There are times, however, when the so-called immediacy behavior of eye contact leads to negative feelings and attitude inferences (Mehrabian, 1969b). For example, the gaze of another can make us squirm uncomfortably in our seats and/or provoke us into prolonging our return gaze in angry defiance. Evil Eye beliefs are widespread evidence of negative experiences with direct gaze. Many of us

**TABLE 1-1 (Continued)**

Body cues (*continued*)
   Movement (facial and
      other body parts)
     Mobile
     Immobile

| Gestures of hands, legs, arms | |
|---|---|
| Body orientation | Body movement |
| Toward | Swift |
| Away | Smooth |
| Downward | |
| General appearance | |
| Body size | Clothing |
| Height | Color |
| Weight | Style |
| Age | General facial characteristics |
| Wrinkles | Eyes |
| Hair color | Nose |
| Voice cues | Mouth |
| Pitch | Tone |
| Loudness | Tempo |

Interpersonal space (definitions)
   Area immediately surrounding individual in which most actions take place
     (Little, 1965)
   Potential space within which one can hold or touch other (Watson &
     Graves, 1966)
   Axial space—between outermost limit of limbs (Machotka, 1965)
   Distal space—within the limits of the distance receptors (Machotka, 1965)

are also acquainted with "paranoid fantasies" involving "accusing, staring eyes" (Freedman, 1979). Such commonplace associations with eyes, along with evidence that eye contact is used aggressively among higher primates, make it no surprise that psychological researchers have found that returning a starer's gaze can be, under certain circumstances, an aversive experience (Ellsworth & Carlsmith, 1973). While Rubin (1970) documented a positive relationship between eye contact and loving, others have shown that direct gaze is sometimes seen as an act of dominance (Strongman & Champness, 1968; also see Chapter IV). It has also been found to be an indication of dislike, especially when combined with a hostile facial expression (Argyle, 1973).

Staring has been found to be related to an increase in the interaction distance of stare recipients (Argyle & Dean, 1965) and has also been shown to increase the speed of departure (Ellsworth, Carlsmith, & Henson, 1972; Elman, Schulte, & Bukoff, 1977; Greenbaum, 1977). Alice Dan (cited in Freedman, 1979) studied the effects of observers (two male and one female, herself) staring at persons eating alone in a lunchroom. Most interesting were the experimenters' descriptions of how it felt when the subject returned the observer's gaze:

> When the observer's direct gaze was returned by a subject, there was invariably a feeling of tension reported by all observers. In two cases the tension was dissipated by a subject's smile, but more often it was broken when one of the starers looked away.
>
> Male observer No. 2 reported that when a male subject stared back, "My heart beat faster, I felt adrenalin going all over the place. I felt he was giving me a sinister stare. I felt the same as if I knew he were going to hit me. (Freedman, 1979, p. 77)

As indicated earlier, eye contact often acts to intensify the feelings of an interpersonal exchange. "If what a person says is friendly or approving, eye contact will make it seem even friendlier. If what a person says is unfriendly or disapproving, . . . eye contact will serve to emphasize the hostility" (Rubin, 1973, p. 35). Subjects who were evaluated by a hostile interviewer rated the situation and the interviewer more negatively when the interviewer made frequent eye contact than when eye contact was infrequent (Ellsworth & Carlsmith, 1968). Such subjects also avoided looking at the hostile interviewer, thus behaving in a manner similar to that of submissive primates (Exline & Winters, 1965b).

Ellsworth, Carlsmith, and Henson (1972) conducted a series of field experiments which tested the hypotheses that staring elicits avoidance behavior in humans. Experimenters on motor scooters or standing on a street corner either stared or did not stare at pedestrians or automobile drivers. The recipients of the stare crossed the street significantly faster than those who did not receive a stare. The experimenters offered two possible explanations for these results. The first was that "staring is generally perceived as a signal of hostile intent, as in primates, and elicits avoidance . . ." (p. 310). The second was that "gazing at a person's face is an extremely salient stimulus with implications which cannot be ignored. The stare, in effect, is a demand for a response, and in a situation where there is no appropriate response, tension will be evoked, and the subject will be motivated to escape the situation" (p. 310). Ellsworth (1975) later dismissed the idea that a stare is always and innately a threat and emphasized the importance of social context in the labeling of the arousal created by mutual gaze.

Testing Ellsworth's notion, Ellsworth and Langer (1976) set up experimental distress situations which called for a bystander's help and which varied according to level of ambiguity. The experimenters hypothesized that when someone received an explicit message regarding a distressed individual's needs, eye contact would elicit an assistance response, whereas an ambiguous message combined with eye contact would make the bystander uncomfortable and lead to avoidance. Ambiguity was varied by having a confederate either inform the bystander that a person behind the subject needed assistance in finding a contact lens (no ambiguity situation) or that a person looked ill and was in need of help (high ambiguity situation). The experimental hypothesis was proven and supported Ellsworth's argument that "staring is not necessarily perceived as a threatening signal, and does not automatically elicit flight" (1975, p. 18). The nature of the circumstances surrounding the gaze determined the approach or avoidance effects of its arousal. (Of course this piece of research is a perfect example of how one must be careful not to generalize on the basis of some variables and not to account for others. It is hard to compare the distress of losing a contact lens with that of being ill. It is not clear here that "ambiguity" is the only distinguishing variable.)

Generally speaking, whether or not one likes another person who makes eye contact depends on whether the interaction is positive or negative (Ellsworth & Carlsmith, 1968), threatening or comfortable (Ellsworth & Ross, 1975). Another situational factor which might influence a response to eye contact is the nature of one's last experience with looking behavior (Tankard, 1970). This possibility is indicated by Tankard's (1970) finding that the sequential order in which eye positions were viewed affected subjects' responses to change in the eye positions.

Although the experimental literature has shown that complex situational and mediating cognitive factors must be taken into account in order to identify the "meaning" of eye contact (Harper et al., 1978), many people make snap judgments about others based on eye contact alone. The mere occurrence or nonoccurrence of eye contact is one of the factors which influence how one person will respond to another (Tankard, 1970). The impressions we form about others' temporary and permanent states of being are often based on their visual behavior (Argyle & Cook, 1976).

When based on eye activity, the accuracy of people's ideas about others varies. The direction and amount of gaze are often used as evidence for certain stereotyped attributions, and research has shown that people are reasonably accurate when it comes to gaze stereotypes about liking and hostility (Argyle & Cook, 1976). On the other hand, people are likely to assume that an averted gaze indicates a person is anxious or tense (Cook & Smith, 1975; Kleck & Nuessle, 1968). There is, in fact, only equivocal evidence for such a correlation (Argyle & Cook, 1976). And while many suppose that powerful people gaze more (Cook & Smith, 1975; Argyle, Lefebre, & Cook, 1974; Kleck & Nuessle, 1968), data indicates that in reality they often gaze less (Exline, 1971; see also Chapters II and III).

Perhaps if people were more aware of the contextual cues surrounding eye behavior, their judgments of the meaning of this behavior would be more accurate. One of the important factors to keep in mind when we interpret the eye language of others is their cultural background. The following section explores the "nature versus nurture" controversy regarding the use and meaning of eye behavior in communication. Cross-cultural comparisons reveal that we cannot assume that patterns of eye contact are the same for those who come from cultures different from our own.


## CULTURAL DIFFERENCES IN EYE LANGUAGE

In 1872 it was suggested by Charles Darwin in *The Expression of Emotions in Man and Animals* that facial expressions of emotion are similar among humans, regardless of culture, because of their evolutionary origin. In agreement with Darwin's hypothesis, Hart (1949) asserted that expressive eye movements are instinctual and not learned. In 1969, Ekman and his associates found empirical evidence to support Darwin's hypothesis through studies in New Guinea, Borneo, the United States, Brazil, and Japan. Observers in these cultures consistently related the affects of happiness, fear, anger, disgust, contempt, and sadness to specific facial patterns. These facial expressions also have been found in infants and in deaf and blind children (Argyle, 1975), and some of them are very similar to those of primate expressions, to whom their evolutionary development has been traced (Chevalier-Skolnikoff, 1973). Also universal is the fact that the same kinds of information are communicated in the nonverbal mode, i.e., interpersonal attitudes, emotions, and other information about the self. Nonverbal communication is also a universally integral part of speech, art, and ritual (Argyle, 1975). Eibl-Eibesfeldt (1972) discovered a friendly flirting gesture which was common to peoples as diverse as the Balinese, Papuans, French, and Waika Indians. (This facial gesture is composed of a rapid raising and lowering of the eyebrows accompanied by a smile and often by a nod.)

Darwin's theory has met with some disagreement and qualification, however. A debate over whether facial expression of emotion is learned, and therefore culture-specific, or whether it is determined by evolution, and therefore universal, surfaced in the psychological literature on nonverbal expression. However, in 1969, Ekman and Friesen clarified the issue with what was later called their "neurocultural theory" of facial expression (Ekman, 1972). They proposed that neurology forms the basis for the universality of emotional facial expression (i.e., the fact that people everywhere express happiness, fear, anger, disgust, contempt, and sadness with the same facial configuration), but that cultural rules govern the propriety of facial expressions, thus allowing for the development of differences in facial behaviors between cultures (Friesen, 1972). Such differences are determined by the fact that the events which elicit certain emotions may vary from culture to culture. Furthermore, cultural "display rules" may cause a person deliberately to interfere with or mask the expression of emotion. It appears, therefore, that a person's style of nonverbal expression will not only reflect the idiosyncracies peculiar to that individual, but it will also contain certain innate, universal elements and follow the unique code of nonverbal expression shared within her/his own culture.

Argyle (1975) has proposed that cultural differences in bodily communication express what he calls the "broader aspects of national character" (pp. 86–87). The nonverbal cultural code which evolves in each culture can be seen as relating to the underlying world view and value system unique to that culture—just as individual styles of nonverbal behavior are related to the person's underlying personality structure and motivations (see Chapter III). For example, Argyle (1975) hypothesized that:

> in cultures where extraversion or affiliative motivation are high (U.S.A., Australia) we would expect more proximity or more gaze, or other signals related to intimacy. In cultures in which "face" and self-esteem are very important (Japan) we would expect more self-presentation behavior, and greater control over the information emitted about the self. (p. 87).

Although more research is needed in this area, it seems that both "national character" and nonverbal behavior are influenced by shared socialization experiences within individual cultures (Argyle, 1975). For example, when babies are carried on a parent's back, the children may spend less time looking into the parent's face and may thus be less inclined to look at faces later (Argyle, 1975). This custom may help to explain the relatively greater gaze aversion found among the Japanese. Among the Sebei in Southern Uganda, mothers have been observed to make little eye contact with their infants. An examination of photographs (Goldschmidt, 1975) showed eye contact occurred in only 1 out of 27 instances. Goldschmidt (1975) interpreted these data to reveal a low affect level in mother–child relationships and proposed that such distant nonverbal behavior between mother and child results in the transmission of emotional noninvolvement, a trait which appears to be characteristic of Sebei adults.

Hall (1963) proposed the categories of "contact" or "noncontact" to describe characteristic proxemic patterns of different cultures. Relatively small interpersonal space, frequent touching, more direct axis of orientation, and direct eye gaze were considered characteristic of "contact" cultures. Although Watson and Graves (1966) and Watson (1970) confirmed the contact/noncontact dichotomy through systematic measurement of the proxemic patterns of many national groups, Noesjirwan (1978) found that Hall's dichotomy was probably an oversimplification, due partly to methodological problems.

Watson and his associates' findings (1966) were based on foreign students within the United States. The prevailing rules of U.S. proxemic patterns no doubt influenced the nonverbal behavior of "the sojourner in a strange culture" (Noesjirwan, 1978, p. 334). Whereas Watson (1970) had categorized Indonesians as a noncontact culture, Noesjirwan (1978) found that they actually used a smaller interpersonal distance than the Australians, who had been categorized previously as high-contact. Rather than contact/noncontact, Noesjirwan (1978) proposed the more useful concepts of affiliation and dominance:

> While affiliation appears to be signalled by closeness, touching, and smiling, dominance appears to be signalled partly by direct body orientation and eye gaze (Strongman & Champness, 1968; Cook, 1970). If this conceptualization has cross-cultural validity, then it may be argued that Indonesian culture, in comparison to Australian culture, is characterized by higher affiliative patterns and higher deference (lower dominance) patterns. (p. 334)

Although it is now apparent that patterns of body language vary between cultures much as spoken language does, it has only been fairly recently that we have recognized the significance of such differences in nonverbal communication (Hall, 1959). There has been little research on nonverbal cultural differences in general, and even less on cultural variations in mutual gaze. However, since eye contact is considered a potent interaction involving intimacy, it would follow that as cultures create rules governing the amount of intimacy to be tolerated, so would they create rules determining how much and what kinds of eye contact are permissible (Tomkins, 1963).

Each culture specifies "at what, at whom, and how one looks, as well as the amount of communication that takes place via the eye" (Hall, 1963, p. 1012). A "moral" looking time is generally expected within each culture. In the United States, for example, if a person looks "too long", s/he suggests intimacy or a power struggle; if s/he looks too little, s/he suggests disinterest, dishonesty, or suspicion (Ekman & Friesen, 1969).

Watson (1970) showed that Arabs, Latin Americans, and southern Europeans focused their gaze on the eyes or face of their conversational partner. It has also been found that a pair of Arabs engaged in conversation with each other in the laboratory looked at each other more than two United States Americans or two English people (Argyle, 1975). Arabs also looked more even though they generally stood closer together. Eye contact is very important to Arab people, which makes it difficult for them to talk with someone wearing dark glasses or when walking side by side (Argyle, 1975). According to Argyle (1975), the Arabs are very sensitive to nonverbal communication because much of their behavior is regulated by social formalities and their language is "stereotyped."

In Japan, nonverbal communication is inhibited in the expression of emotions, but is very important in signalling social status and the nature of relations between people. Similar to other Asians, such as Indians and Pakistanis, and to northern Europeans, the Japanese orient themselves toward others without looking directly into the face or eyes (Watson, 1970). They are taught to look at the neck and they look at the faces of superiors as little as possible (Argyle, 1975). In Japan, eye contact with strangers in public places is generally avoided, but in mass transit vehicles people will briefly scan the others present, turning their heads from side to side (Argyle, 1975). The Japanese inhibit facial expression of emotion and thus pay less attention to faces. According to Argyle (1975), in Japan the "ideal is a controlled, expressionless face, and laughter or smiling may be used to conceal anger or grief" (p. 75). There is more emphasis on other kinds of nonverbal communication in Japan.

Although there are few formal studies of cultural variations in eye contact, the problems of those who visit or move to different cultures provide evidence that rules regarding eye contact behavior do indeed differ across cultures. A woman who moved from the Far East to the

United States at the age of twenty-five found the issue of eye contact disturbing enough to write to "Miss Manners," an etiquette columnist in *The Washington Post.* The woman wrote that, in the Far East, "staring into the pupils of another is the height of impropriety (or come-and-get-it-ness) . . . my adjustment to the United States will never be complete without some kind of 'code of looks'" (Martin, 1981, p. B5). As "Miss Manners" rightly pointed out, it is perhaps more important to learn these rules of "eye etiquette" than other rules "because people often fail to realize that such behavior as eye contact is learned, and they pounce on it as being psychologically revealing" (p. B5). "Miss Manners" went on to advise: "In this country, it is considered polite to look people in the eye when conversing with them. If you find this difficult, an alternative is to look away but maintain a smile, in which case your behavior will be misinterpreted as charming demureness. That, at least, is better than being considered shifty." "Miss Manners" might well consult with research in order to give more precise advice.

At the Nonverbal Communication Research Conference on Body Politics (1980), Judith NineCurt described the angry stares that she, as a Puerto Rican woman, received from mainland United States women because her pattern of eye contact was considered an invasion and thus highly insulting. While passing women her age on the street, she would look at them and smile before looking away. This pattern was "normal" in her town in Puerto Rico, but on the mainland it was received with hostility. NineCurt noted that when talking with someone in Puerto Rico you look straight at the person, but not as directly as on the mainland United States. She described that in Puerto Rico the head swings from side to side, in contrast to the more penetrating "tunnel gaze" of the mainland United States look. As a result of this and other cultural differences, NineCurt said she experienced much loneliness and culture shock.

When we interact with members of our own culture, the similarity of expressive behavior provides a "background of rapport" (Dabbs, 1969) which is most often taken for granted. It can be quite disturbing, however, when someone uses what we consider to be an "inappropriate" nonverbal response (an experience similar to hearing a word pronounced in an unfamiliar manner). Because our body movements are not random but are synchronized with our speech, the nonverbal difficulties of communication in another "culture, particularly where the language and language rhythms are different from our own, should be far greater than ever appreciated. Not only must we learn to hear and express the new sounds in new rhythms of another language, we must also learn to blink and twitch in a rhythm" (Condon & Yousef, 1975, p. 127).

Another complicating factor when we are conversing in a second language is that we may rely even more on nonverbal behavior for its communicative value than we would when conversing in our native tongue. In such situations, violations of nonverbal norms become more glaring. Specialists in intercultural communication have observed that "we much more readily expect, excuse, or correct errors in spoken language than errors in nonverbal behavior" (Condon & Yousef, 1975, p. 259) and it is extremely awkward to ask for clarification of a bodily movement.

Emotions, the primary content of much nonverbal communication, are particularly difficult to communicate between persons of different cultures because of problems of misinterpretation. Differences in nonverbal expression add fuel to ethnic stereotypes and racist beliefs because although two persons from different cultures may converse in the same language, they perceive each other as unable to make "genuine" or "appropriate" emotional responses. Both persons involved in such interactions may feel the other is either without feeling or exhibiting "phony" feelings.

Thus, misinterpretation due to cultural differences can lead to alienation and culture shock; it can contribute to desynchronized conversation and thus, general discomfort; and it can fuel stereotypical interpersonal judgments. It is encouraging to note that there has been some investigation of the possibility for improving intercultural relations through the training of nonverbal behavior. Collett (1971) trained Englishmen in the nonverbal behavior of Arabs, which included a high level of direct gaze. An Englishman who used Arab nonverbal displays was preferred over one who did not by the male Arab participants (all participants were males). It is hoped that researchers will test further the possibilities for cross-cultural nonverbal training and that such knowledge will be applied toward the goal of world harmony.

## CONCLUSION

Having sifted through the psychological data regarding the language of eye contact, we can use the following general guidelines to increase our command of the language:

If you cannot say it with words or with words alone, try saying it nonverbally. Use your eyes to get others' attention and engage them in conversation. Likewise, to avoid interacting with someone, avoid her/ his eyes; to avoid interruptions at work, keep your back to the door.

If there is a breakdown in the flow of conversation between yourself and another, it could be because your partner-in-conversation and yourself have different patterns of eye contact. In the United States and Britain, people who have been observed in psychological studies (such experimental subjects are from predominately white college student populations) periodically gaze into each others' eyes while conversing. Typically, eye contact occurs during ten to thirty percent of the conversation, with mutual gazes lasting about one second each. The people studied tend to look more and for longer periods while listening and to look least while talking. Looks between utterances help cue who will speak next or continue speaking.

Interactants who have different cultural backgrounds and thus different patterns of eye contact will have trouble synchronizing their conversational behaviors. Also, when we cannot see the eyes of others, as in telephone conversations or when one interactant wears dark or reflective sunglasses, we may also expect more interruptions and missed cues which disrupt smooth communication because the eyes cannot be used to signal turn-taking. In such situations a nonverbal feedback loop is broken and communication breaks down.

Satisfying communication is best achieved when both interactants use the same modes of communication. We are more likely to feel that we have truly communicated if our partner speaks and gazes intermittently in a pattern similar to our own. Likewise, we can foster a feeling of satisfying communication in our partners by responding to them with our eyes as they use theirs.

If the eyes are not "windows of the soul," at the very least they are portraits of an individual's emotional state. The eyes serve as a canvas for the vast range of human emotional expression, grinning with joy, awash with tears, gaping at the terrible, hiding with embarrassed fringed shades, or narrowing into anger's focused lens. However, the meaning of a person's look into your eyes can appear ambiguous if the person is not obviously emoting. If so, we can deduce the meaning of their attention by taking into account their other nonverbal behaviors, their verbal behavior, and numerous other contextual factors which cue us as to the meaning of the look.

When it comes to communicating our feelings accurately to others, one must remember that eye contact among strangers is usually perceived differently than eye contact among friends. Your pattern of eye contact may be perceived as "intrusive" or "affiliative" depending on the situation you are in and with whom you are communicating.

# NOTES

[1]The distance between bodies in focused interactions has been shown to vary according to the nature of the interaction. Hall (1979b) distinguished between four different "distance sets": intimate, personal, social consultative, and public. Different senses operate in the different sets. For example, at an "intimate" distance, touch, olfaction, and thermal senses are primarily employed, while the sense of sight plays a relatively minor role. At the limit of the "personal" distance, where participants are about five feet apart, vision and hearing are the only senses used (Argyle & Kendon, 1967). People in the United States generally choose a nose-to-nose distance of approximately five feet (Sommer, 1959). Standing two or three feet apart is considered too close (Lassen, 1973; Patterson & Sechrest, 1970). Friends are expected to stand closer to one another than strangers or acquaintances, and distances between people tend to be smaller in impersonal environments (such as a public street) than in more intimate environments (such as a living room) (Little, 1965). Body orientation varies according to the manner in which it facilitates or inhibits eye contact (Sommer, 1959), as well as according to the kind of interaction (Hall, 1963). People who are casually conversing at a table often sit diagonally across the corner from each other rather than side by side or directly opposite (Sommer, 1959). However, in a competitive interaction, individuals will tend to sit directly opposite one another (Sommer, 1959).

[2]Unfortunately, such laboratory studies of gaze and physiological arousal cannot be accepted as conclusive. More research is needed to clarify whether their results can be generalized to everyday situations (Fehr & Exline, in press). In the majority of the above studies, participants gazed continuously in silence at confederates who returned gaze at prearranged schedules. It has been observed that continuous mutual gazes of 10–18 seconds in the absence of conversation are nonnormative (Donovan & Leavitt, 1980) and may be arousing due to the aversiveness of the stare in silence (also discussed later in this chapter).

[3]A later study by Cary (1979) indicated that Goffman's (1963) anecdotal observations of "civil inattention" did not hold true for persons passing one another on a college walkway. The most typical nonverbal display between passing persons was composed of holding head erect and gazing straight forward (rather than at least glancing initially toward others, as Goffman would have predicted). Fehr and Exline (in press) noted that this is "a variety of inattention, perhaps without the 'civility' " (p. 54).

[4]Interacting persons often closely coordinate their movements with each other (Condon & Ogston, 1966). Scheflen (1963, 1964, 1966, 1973, 1974) has observed that the phenomenon of "postural congruence" occurs among people who know each other and are involved in a shared objective. (The postures of two or more interactants are congruent when they hold their heads and extremities—arms, hands, legs, and feet—in the same position).

Postural noncongruence indicates a person's lack of association with an interaction and is often accompanied by other disconnecting behaviors, such as avoidance of eye contact (Trout & Rosenfeld, 1980).

[5]In such experiments, confederates were instructed to gaze continuously at subjects. The subject's eye contact with the confederate was recorded by observers who were looking directly into the subject's eyes from behind a one-way mirror (located behind the confederate). In order to avoid the distraction of subjects seeing their own reflections, they were placed at an angle to the one-way mirror. Because the confederate stared continuously, the total amount of eye contact was the same as the amount of time that the subject looked at the confederate in the region of the eyes, which was recorded on a cumulative stop watch. An alternative to this kind of laboratory setup is the filming or videotaping of interactions between two "genuine" interacting subjects (e.g., Kendon, 1967).

[6]Some researchers have found evidence which they believe links eye color and personal behavior (Buffington, 1984). For example, a study which compared light- and dark-eyed children found "dramatic differences" in their reactions. Dark-eyed children worked better under pressure, if they were given the opportunity to exercise their own initiative in solving problems. Light-eyed children "worked better after careful consideration, application and stick-to-itiveness. These children were also more widely read and better able to remember information than their dark-eyed cohorts" (Buffington, 1984, p. 93). More research in this area is necessary, however, before definitive conclusions can be drawn.

[7]In China, jade dealers determine a potential buyer's interest by observing pupil dilation, so astute shoppers wear sunglasses (Webb, Campbell, Schwartz, & Sechrest, 1966). In Arab culture it is common knowledge that pupils dilate when you are interested in something, and contract when someone says something you do not like (Hall, 1979). Since these responses of the eyes cannot be controlled, many Arabs wear dark glasses, even indoors. Market researchers rely on this aspect of eye language to get consumer reaction to new products and advertisements when they record the size of the dilation response ("And Now a Word About Commercials," 1968; Wahl, 1969).

[8]Mehrabian (1969a) defined immediacy "in a somewhat general form as the extent to which communication behaviors enhance closeness to and nonverbal interaction with another" (p. 203).

# CHAPTER II

# Eye Contact and Relationships of Power: Charisma and Dominance

In the previous chapter we have seen the vital function of eye contact in social interaction as it serves to regulate or coordinate the behavior of two participants in the joint activity of a conversation. We have also seen how eye contact functions as a mechanism for emotional communication.

As noted in the closing section of Chapter I, patterns of eye contact are affected by the unique cultural milieu in which they have been developed. Within any society, there are assumptions concerning eye behavior, such as who may look at whom and when looking may occur. The present chapter and the next are concerned with how such rules for eye behavior reflect—and maintain—the social and political relationships of power within our society. The interplay of eye contact between two people can tell us not only their attitudes and emotions, but also their relative power positions and dominance.

In this chapter we will explore the function of eye contact as a mechanism of power and examine visual behavior as a reflection of the social arena. The power of the eyes to exert special influence and the use of eye contact by certain individuals to manipulate interactions are discussed in the first section. In the next section, the interrelation of power roles and visual behavior is described and the specific behaviors of dominance and submission are compared.

Most of the popular and empirical literature on nonverbal behavior focuses on the emotional and solidarity aspects of the interaction (Henley, 1973–1974). The communicators are assumed to be "equal" and the primary concern is whether, how, or why a particular nonverbal behavior affects a given interaction; for example, whether eye contact increases liking, how eye contact affects a cooperative or competitive joint activity, or why romantic partners look more at each other. The literature on eye contact emphasizes relations of friendship, affiliation, or intimacy, rather than relations based on power, status, and social control. We know that eye contact carries a social message and that visual encounters are important for cuing the interactants to one another's intentions. What the literature generally fails to recognize is that these cues are particularly significant in establishing or maintaining relationships of power.

Power may be defined as the ability to determine or influence the outcomes of situations. A relationship of power is inherently unequal, and the "ability" of the powerful is based upon their control of resources. Access to resources, much less control of them, is largely determined by the macropolitical structure of a society, through its institutions and customs. If we think of power as being exercised along a continuum, we can see that control ranges from the subtle (such as socializing females to avert their eyes when men stare at them) to the overt, as exemplified by the lynching of Emmett Till, a fifteen-year-old black youth, for staring at a white woman. Nonverbal communication can be located along this continuum at the point between covert and overt control (Henley, 1976). The subtle yet quite visible signals of nonverbal communication occur in social interactions, thus serving as the most frequent means for social control. Messages about power are exchanged in thousands of daily behaviors through which nonverbal influence occurs, and which ultimately support and maintain the macropolitical structure.

When we put nonverbal behavior into its political context, many nonverbal acts may be seen as dominance signals by the powerful and submission signals by the powerless. The behavior of the eyes, in particular, reveals the balance of power: persons of greater power stare more at others, look less as they listen, and command a larger visual "space," while persons with less power avert their gaze when stared at, look more attentively as they listen, and assume less visual space. The relationship between these behaviors and the sociopolitical structure becomes clear as we examine who is most likely to exhibit these respective patterns—the executive or the janitor, men or women, blacks or whites (see Chapter III).

## EYE CONTACT AND PERSONAL POWER

Anyone can play a power game with the eyes. Simply stare at a stranger for more than a few seconds. Once the stranger is aware of your staring, s/he will probably show signs of discomfort, such as shifting her/his body or even leaving the area. But beware! Your staring may also produce a challenge, especially if your eyes lock with another pair of increasingly angry eyes.

Of course, most of us do not deliberately play these kinds of "power games" with the eyes, but most of us are aware that sometimes we have used eye contact, or seen someone use it, to manipulate a social situation and get what s/he wants. Indeed, psychological studies have found that looking into a person's eyes while s/he is performing a certain behavior will increase the chances that s/he will do it again.[1] It has also been found that eye contact increases the likelihood that those we petition will comply with our request (if the request is perceived as "legitimate").[2]

We are familiar with people who have used their eye contact like a magnet, to draw in, captivate, and/or manipulate others. This power of the eyes is humorously illustrated by the character of Evil-Eye Fleegle, in the popular comic strip "Li'l Abner" (Guthrie, 1976). Evil-Eye uses a "whammy" or even a "double whammy" to zap his opponents. The "whammy" is a type of stare in which the eyes become very large and transmit concentric rings of power towards the victim.

While people do not have the capacity to "whammy" each other, the effect on the receiver of certain kinds of eye contact may be similar. Some politicians, religious and cult figures, "super" salespeople, criminal "cons", even musical conductors and therapists sometimes possess a dynamism which is manifest through their eyes. Their gaze exerts a strong, charismatic influence on others. This influence is evident in the following anecdote about Lawrence of Arabia, told to Lowell Thomas:

> One morning I was in the tent with Lawrence when a young Bedouin was hauled in charged with having the evil eye. . . . Lawrence ordered the offender to sit at the opposite side of the tent and look at him. Then for ten minutes he regarded him with a steady gaze, his steel-blue eyes seeming to bore a hole into the culprit's very soul. At the end of the ten minutes, Lawrence dismissed the Bedouin. The evil spell had been driven off! (Guthrie, 1976, p. 118)

It was probably the combination of Lawrence's steel-blue eyes, and his dynamic personality and position of power, that "drove off" the spell of the evil eye. Using the eyes to manipulate the feelings or behavior of others is a common characteristic of the powerful and charismatic. The manipulator can generate a range of emotion in the other, from helplessness to servile devotion. S/he may be a political leader, a cult figure, or a professional of some sort. For example, one client described her former therapist as a man who spent long periods of time meditating with her and gazing into her eyes as she revealed her feelings. She said she felt "mesmerized" by her therapist and vulnerable to his power. Although she believed that he had destructive power over her, it was extremely difficult for her to break away from him.

This technique of therapeutic domination is not new. Shamans, "medicine men," and "voodoo doctors" often relied on their eye contact to practice their particular forms of "therapy." Charles VI of France was supposedly dealt with in a similar manner, for intermittent periods of madness, by an "evil-eyed charlatan" (Tuchman, 1978). The reputation of the notorious Russian monk, Rasputin, was built around this dynamic eye contact and hypnotic power. Rasputin gained control over the imperial couple through his miraculous "cure" of their hemophiliac son.

> (He) told me to lie down on the couch. He stood in front of me, looked me intently in the eyes. . . . His hypnotic power was immense. I felt it subduing me and diffusing warmth throughout the whole of my being. I grew numb; my body seemed paralyzed. I tried to speak, but my tongue would not obey me, and I seemed to be falling asleep, as if under the influence of a strong narcotic. Yet Rasputin's eyes shone before me like a kind of phosphorescent light. From them came two rays that flowed into each other and merged into a glowing circle. This circle now moved away from me, now came nearer and nearer. When it approached me it seemed as if I began to distinguish his eyes, but at that very moment, they would vanish into the circle which then moved further away. (Wilson, 1974, pp. 104–105)

Certainly Rasputin was not universally loved; he was thought by many to be a charlatan, a con man, and a spiritual fraud. Upon meeting him, some people were noticeably unimpressed: "He ran his pale eyes over me, mumbled mysterious and inarticulate words from the Scriptures, made strange movements with his hands, and I began to feel an indescribable loathing for the vermin sitting opposite me" (Wilson, 1964, p. 91). Yet it seems likely, given all accounts, that Rasputin did have some hypnotic or healing power and that he used it effectively with a number of people (Minney, 1973).

Leaders of various religious sects and cults have been known historically for their dynamic or penetrating eyes.* (See below.) Some other contemporary groups and cults, all of which have been compared to brainwashing, use extensive eye contact with strangers to provide a kind of instant intimacy. For example, at an introductory meeting,

---

*Most recently the cult leader Bhagwan Shree Rajneesh has been described as a "doe-eyed leader" ("Busting the Bhagwan," Nov. 11, 1985, p. 26). The news media suggested that his eyes are powerful.

members of a particular cult-like group are trained to walk up to a stranger in the group, quickly scan the name tag, and look deeply into their eyes while greeting the person by name. The combination of eye contact and being called by name throws one into an immediate encounter with the cult member. If one does not feel manipulated, one may feel "cared about," "important," "noticed," or "recognized," instead of alienated or alone. This technique has been used also in other contexts to break down a person's defenses and make one more vulnerable to indoctrination.

The comic strip, "On Stage" (Starr, 1979), depicted a cult leader who was strikingly blue-eyed and who wore a large blue eye in the center of his forehead (the third eye, which is associated with increased spiritual sensitivity). The cult disciples wore a similar "third eye" medallion around their necks.

The use of sunglasses by Jim Jones, leader of the People's Temple cult, prevented others from meeting his gaze while at the same time allowing him to stare into their eyes. This simple strategy of using a one-way process of receiving signals without sending any helped to establish his dominant position. The lack of social information leads to uncertainties about how to respond, putting the sunglasses wearer in a more powerful position. Guthrie (1976) called this source of intimidation the "shades effect" and described the annoying discomfort of talking with someone who is wearing glasses that obscure the signals being transmitted.

The loss of personal identity and/or a change of mannerisms is common in destructive cults. Though the members who recruit others to a cult may use intensive eye contact to intimidate or establish intimacy with the newcomer, other cult members may be characterized by a "peculiar smile or staring expression . . . some may appear as if in a trance-like state. . . ." (Shapiro, 1977, p. 83). The lack of animation in the eyes of these cult members may reflect the self-control they have surrendered to the cult or its leader(s).

The power of the eyes has long been associated with the capacity to exert special influence on a subject or victim, especially within a restrictive context, such as a cult, a therapeutic relationship, or a closed society of some sort.

"Fascination" at one time held a special place among hypnotic phenomena, being ascribed to some mysterious emanations from the eyes and being said to be the means by which some wild beasts overpower their prey. The snake, in particular, was thought to have great powers of fascination and there are tales of these reptiles forcing birds perched on the branches of trees to fall into their open jaws by merely looking at them. In Europe and other parts of the world there used to be (and probably still are) beliefs that certain "thieves of women and children" made a practice of kidnapping by merely looking victims in the eyes and fascinating them into following wherever they lead. . . . It is conceivable that in a culture where such a belief was prevalent and taught to its members from infancy, such fascination would be a reality, a product of social suggestion. (Weitzenhoffer, 1957, p. 118)

The fear of being enveloped or devoured by someone's eyes, and a concomitant fear of having one's control surrendered to another, probably accounts for the reluctance of many people to be hypnotized (see Chapter IV), even by a professional who has explained the procedure. For over two hundred years the magical aura surrounding mesmerism, later known as hypnotism, resulted in the widespread belief that a subject could not exert her/his will and voluntarily break out of hypnosis at any time. The "Svengali myth" of one person controlling another's will by eyes that stare into their soul is quite prevalent.

Hypnosis, as a technique, has come a long way since Franz Mesmer, an Austrian physician, developed a method of using the eyes to cure afflictions. Mesmer induced trances by gazing into the eyes of his outpatients and putting his hands on their bodies (Brenman & Gill, 1971; Hartland, 1971). While in a trance, the mesmerized subject would usually dance, gyrate, or move uncontrollably until a fainting spell occurred. Upon awakening, she would be "cured" of the affliction. Mesmer claimed that a magnetic power caused the trance, and that a "divine power" could be transmitted to an ailing person as a cure. The eyes were the source of this magical cure, and Mesmer would stare directly into the patient's eyes for some minutes (Meares, 1960).

Modern hypnotists know that the real key to the trance is the power of suggestion, and that the subject is controlled by the hypnotist only to the extent that s/he believes s/he is controlled. Nevertheless, the eyes continue to play a major role in the process. Scharff (1978) outlined four contemporary means of hypnotic induction, all of which in some way involve the eyes: direct eye gaze, eye fixation, eye catalepsy, and the shock method.

The direct gaze method involves catching and holding the subject's attention as s/he stares into the hypnotist's eyes. The hypnotist sits close to the subject, whose eyes eventually tire and close. The subject usually believes that the hypnotist's stare causes this feeling of fatigue (LeCron & Bordeaux, 1947). Although the method induces a deeper hypnotic state, the hypnotist is also prey to the powerful gaze of the subject, and on rare occasions, the direct eye gaze method may result in both of them becoming immobilized! This technique has fallen into comparative disuse since the recent trend toward a more passive induction (Meares, 1960), that is, using relaxing imagery with the eyes closed.

The direct eye gaze method is also considered to be an authoritarian technique. Yet Meares (1960), LeCron and Bordeaux (1947), and others claim that the direct eye gaze is a superior method of hypnotic induction. As noted in Chapter I, Meares (1960) contended that the reason staring at the hypnotist's eyes facilitates a more rapid trance than, say, looking at a bright object is because of the cultural, symbolic, mythological, and folkloric significance of the eyes. He suggested that the direct stare in hypnotism conjures up either the eye of authority or an all-seeing eye.

A second commonly used method of induction is eye fixation. Combined with verbal suggestions of sleepiness, the subject fixates his/her eyes on a spot on the ceiling, a bright object, a moving pendulum, or the hypnotist's fingers. After a few minutes the subject's eyes will become tired and close, as in the eye gaze method. It is possible to become hypnotized using this method while driving or looking at television. Because of the dangers involved with this kind of "self-hypnotism," it is important not to be exposed to the same monotonous configuration for a long period of time (e.g., tired drivers at night may find their eyes fixated on the painted divider lines, rather than watching for other traffic signals).

Eye catalepsy is a third method of induction, defined as the inability to open the eyes. With this technique the hypnotist instructs the subject to look up with the eyes closed and to hold the eyeballs in that position. The subjects may be unable to open their eyes, and may become more suggestible and more deeply hypnotized. In holding this position, it may become physically impossible for the subjects to open their eyelids. It has been found that the further a person is able to roll the eyes upward, the more deeply hypnotized she can become (Spiegel, 1974, 1977). Spiegel (1972) developed the "eye-roll test" to determine what kind of hypnotic subject a person will be according to the ability to roll the eyes upward. Ackerman (1979) hypothesized that "eye-rolling" may be related to an ability to introspect or to a lack of inhibition.

The fourth method of hypnosis, the shock method, is seldom used. It is an approach which can be effective in reaching subjects who are difficult to hypnotize, and works by suddenly flashing a light into the subject's eyes. However, it may have the opposite effect; that is, to confuse or alarm the subject, and therefore use of this method is rare.

Observation of the eyes is important throughout the modern process of hypnosis. The subject's eyes can communicate emotions such as fear, which the hypnotist must attend to before hypnosis begins (Meares, 1960):

> We want to know everything we possibly can about the moment-to-moment changes in the patient's eyes. We can read anxiety, trust, apprehension, or hypnotic somnolence. If we have him close his eyes or stare at some object, we lose this source of information. Moreover, the patient can use his eyes to communicate ideas which are difficult to put into words: "I am frightened." "Am I all right?" "Don't leave me". . . . If the lids are widely retracted and gaze keeps switching from the therapist to various objects in the room and back again, it is a sign of increasing anxiety, which calls for immediate reassurance. . . . Sometimes the patients close their eyes and keep opening them every few moments. It is often observed that this becomes more and more difficult—a sign, of course, of impending hypnosis. (Meares, 1960, pp. 166–167)

At the start, the subject is told to fix her/his gaze, thus narrowing her/his sensory field, while the hypnotist usually suggests drowsiness and sleep. The hypnotist then watches the eyes to see if they become moist and if the eyelids flutter. When the trance has deepened to an effective point, the hypnotists may tell the subjects that their eyes cannot be opened because their eyelids are "stuck" (LeCron & Bordeaux, 1947).

Several research investigators have confirmed that there are actual physiological changes in the eyes which are a result of the hypnotic state. These include rapid eye movements (Brady & Rosner, 1966), slow eye movement (Weitzenhoffer, 1969), changes in the cornea (Strosberg & Vics, 1962), and changes in pupil size (Bartlett, Faw, & Liebert, 1967). The actual depth of the hypnotic state can be determined by observing the eyes. After a session, the subject's eyes are often glazed, and it continues to be important for the hypnotist to observe the subject's gaze and make sure that s/he is ready to leave the office in a normal state. If the subject has a glassy-eyed appearance, the hypnotist will talk with her/him for a while in order to achieve reorientation.

Just as hypnosis has become an increasingly sophisticated therapeutic technique, so too has our understanding of the eyes as an influential force become more sophisticated. Certainly stage hypnotists are not the only ones who can fascinate an audience. Throughout history, many political figures have been known for the forcefulness of their eye contact. Modern politicians who lack this forcefulness are often schooled in the most advantageous use of their gaze. Willner (1968) identified a number of leaders by their magnetic, irresistible, persuasive eyes. She collected references on the cold, appraising eyes of

Ataturk, the hypnotic eyes of Nasser and Hitler, the X-ray eyes of Lenin, and the profoundly luminous eyes of Mussolini. Sometimes the eyes may be a leader's most striking feature, as in the case of Mussolini, whose "too large, very black piercing eyes . . . seemed almost to protrude from his face" (Kirkpatrick, 1964, p. 156). Another example is Ataturk, whose eyes "gleamed with a cold steady challenging light, forever fixing, observing, reflecting, appraising, moreover uncannily capable of swivelling two ways at once, so that they seemed to see both upwards and downwards, before and behind. With these eyes . . . he had the look of a restless tiger" (Kinross, 1964, p. 163).

Barbara Tuchman, a noted historian, has often characterized important historical figures by utilizing vivid descriptions of their eyes. Many examples are found in her book *The Proud Tower* (1966), where she describes European leaders with eyes that are "flashing," "fixed," "forceful," "fiery," "fishlike," "smouldering," etc. On the modern American political scene, eye contact is as common as speeches in an election year. Political candidates and their assistants are taught to make eye contact with as many people as possible when giving campaign speeches or attending political functions (Sorenson, 1984).

Jimmy Carter is known not only for his ability to smile with a wide grin, but also for being able to make genuine eye contact with many people. Richard Nixon, on the other hand, had to be tutored in certain kinds of nonverbal behavior which did not come naturally for him. His campaign advisors tried to keep him in crowds that were very small, to minimize his problems with making eye contact. They also found it was difficult for him to look directly at a television camera. For this reason, Nixon's speeches were placed directly on top of the camera, so he was forced to make eye contact with it (and thus the viewer). During the 1960 Presidential debates, Nixon's nonverbal discomfort before the TV camera was unfavorably compared with the ease with which John Kennedy faced the camera. A few years later, a popular poster of a particularly shifty-eyed Nixon asked: Would you buy a used car from this man? Ironically, eye contact with Spiro Agnew was rumored to influence Nixon to decide to choose him as a vice-presidential running mate; after looking into Agnew's eyes Nixon said he knew that Agnew was his kind of man (McGinnis, 1969).

The 1980 presidential debates also provided a contrast in nonverbal political and personal style. Jimmy Carter seemed to be staring at the viewer, as if to demand attention to the points of his argument. Ronald Reagan's style was much more low-key; his eye contact is natural and expressive. His practiced ease in looking straight at the camera impressed many people with his "apparent" sincerity. Reagan's acting talent gave him a particular awareness of his behavioral influence, and his ability to project his message clearly, both verbally and nonverbally, made him a potent political figure.

Another candidate in the 1980 election, John Anderson, illustrated another practice of politicians and people in esteemed positions—use of heavy, dark-rimmed glasses. Anderson's glasses gave him an air of directness and purpose by exaggerating his gaze. Many politicians, dictators, business executives, judges, and scientists have chosen similarly dark-rimmed glasses (Guthrie, 1976). When worn by a person in a dominant position, eyeglasses may add to the intimidation of subordinates or rivals in equivalent positions.

Interestingly, preferences for particular political candidates show up in the public's eyes. Barlow (1969) found that pupillary size was a reliable index of preference for a candidate. In his study, subjects who actively supported either a liberal or conservative candidate were shown slides of Lyndon Johnson, Martin Luther King, and George Wallace. The pupils of their right eyes were photographed while viewing the slides. The eyes of the white conservative preference group dilated to the slide of George Wallace and constricted to the slides of Lyndon Johnson and Martin Luther King, while the white and black liberal political preference groups reacted in the opposite manner. Pupillary and verbal preference were in perfect agreement.

Politicians and candidates on the campaign trail are not the only leaders who possess intense gaze or a high level of eye contact. In some cultures, a person is considered a "born leader" because he was born with particular features, such as a penetrating gaze and charismatic personality. The Chinese call this "glitter"—the impression of a person's inner vitality from her/his eyes (Mar, 1974). Many people with aggressive leadership qualities rely on a great amount of eye contact in order to influence or intimidate, or to monitor competition with others. A character in Hesse's Demian advised: "If you want something from someone and you look him firmly in both eyes and he doesn't become ill at ease, give up. You don't have a chance, ever" (1965, p. 59).

Leadership is not only a matter of acting like a leader. More fundamentally, becoming a leader depends upon being recognized as one by others. Burroughs, Schultz, & Autrey (1973) studied the effect of quality of argument on eye contact and the relationship between eye contact and leadership votes. In small discussion groups, the confederate received more eye contact and more leadership votes from naive group members when s/he gave high quality arguments (that is, when s/he displayed leadership). Interestingly, a person can be perceived as a leader without personally displaying any leadership cues (such as expertise or assertiveness) whatsoever. For example, simply taking the position at the head of a table among a group of strangers may be enough to confer leadership status, when other qualifications are held constant (Porter & Geis, 1981).

This "head-of-the-table" effect may be a matter of spatial dominance, since leaders command more physical space than other group members (Goffman, 1959; Henley, 1977). But it may also be a special case of visual dominance, since the person at the head of a table is able to see everybody and does so without having to make any head movements. Everyone else at the table must make some accommodation (e.g., turn the head, lean forward) in order to have eye contact with the leader. Thus, it may be that physical positions at a table reinforce the behavioral and cultural patterns of the leader as visually dominant, and the followers, sitting on the sides, as visually subordinate.

Teachers in training must learn to keep their eyes on their students for, as Hodge (1971) asserted, "at the most basic level, the eyes are an integral part of the teacher's presence in the classroom" (p. 264).[3]

Classroom discipline, one of the biggest problems of the educational system in the United States (Kilpatrick, 1978), can be aided by assertive use of eye contact along with other kinds of body language. Jones (cited in Kilpatrick, 1978) stated that eye contact is very important in conveying that the teacher "means business," and should be combined with quick responses to student behavior and physical proximity. An assertiveness training program for teachers advocated that teachers appear more forceful by making eye contact, placing a hand on a student's shoulder, and using the child's first name in any commands ("How to Get Tough," *Newsweek*, 1978). School time devoted to discipline decreased from 29 to 8% as a result of this program. An experiment involving two elementary school boys, one of whom was in a special education class, showed that verbal reprimands accompanied by eye contact and a firm grasp of the student's shoulders was more effective in reducing disruptive behavior than reprimands without eye contact and firm grasp (Van Houten, Nau, MacKenzie-Keating, & Colanicchia, 1982). Ignoring the disruptive behavior, however, was the least effective method of controlling it.

Obviously the relationships between teachers and students, leaders and followers, politicians and voters, cult figures and disciples, are complex constellations of behavior and interaction. In all of these relationships, a dominant party attempts to influence, persuade, or manipulate the subordinate party. In many of these relationships, eye behavior may be a crucial component in the dominant person's ability to influence, persuade, etc. In this section, we have examined the power of the eyes to exert a particular influence on a subject or victim.

The examples mentioned here have clearly presented themselves as relationships of opposing roles: the one with the most control/power is dominant, while the other is submissive or subordinate. Some of the examples may seem extreme; even if you met a domineering, Rasputin-like leader who mesmerizes his followers through beady-eyed exhortations, it is easy to imagine that you would never fall for such a guy. However, many of us fail to recognize that power roles and relationships are indisputable aspects of our (Rasputin-less) everyday lives.

## DOMINANCE AND SUBMISSION

How nonverbal behavior relates to power roles is an area of study which we have scarcely begun to investigate. We need to know what power-related behaviors are, how they operate, and under what conditions. Some variables have been identified and some patterns are emerging but, on the whole, many questions remain: how are the power roles, dominant and subordinate, reflected or enhanced by nonverbal behavior? How does nonverbal behavior (help to) define or determine a power relationship? Does nonverbal behavior change as one changes one's role? These are difficult questions, but an examination of eye contact and gaze behavior within dominant-submissive relationships will contribute to finding the answers.

Humans are not the only beings who engage in power struggles. We have learned a great deal about the nature of dominance and submission through research on the staring behavior of animals. Dominance is an important variable in animal behavior (Harper, Wiens, & Matarazzo, 1978) and the eyes are the chief parts of the face employed by animals for threatening purposes (Hingston, 1933). The most widely shared aspect of threat behavior in monkeys and apes is the direct stare, and it is often sufficient to establish which animal is more dominant (Altman, 1967; Kummer, 1967; Marler, 1965). The eye contact between male primates initiates mutual threat displays in order to establish one's place in the dominance hierarchy (Exline, 1971). Thus, aggressive staring is most common among the dominant animals in a social hierarchy. Although a subordinate animal sometimes returns the stare temporarily, the characteristic response in a subordinate is gaze aversion and/or flight (Ellsworth, Carlsmith, & Henson, 1972). Gaze aversion therefore is either a sign of submission (Altman, 1967; Andrew, 1963) or an indication of peaceful intent (Ellsworth, Carlsmith, & Henson, 1972).

Some humans have learned the hard way what a direct stare means to an animal. For example, the great horned owl has been known to be so angered by the human stare that, in response, it gouged out its keeper's eyes (Freedman, 1979). In a controlled setting, Exline and Yellin (1969) demonstrated that humans can elicit and inhibit threatening displays of rhesus monkeys simply by initiating and breaking off eye contact with them. They noted that a challenging glance will more frequently elicit attack and threat behavior (e.g., rattling the bars of the cage), while a deferent, averted glance will more likely elicit nonaggressive responses. It appears that an engagement of eyes is a necessary and inherent aspect of the threat stimulus-complex. When the experimenters assumed their stimulus posture with closed eyes the monkeys rarely if ever responded with threatening behavior. Although we cannot yet precisely characterize the context in which eye contact will elicit threat behavior or serve as a dominance signal, it has been suggested that such behaviors are shared by the primate family (Diebold, 1968) and that eye engagement does have an interpersonal regulatory function in the intraspecies relationships of all primates, including humans (Exline, 1971).

Although eye contact or direct gaze is almost always a component of an aggressive encounter between animals, there are some instances in which open staring is nonaggressive. The direct gaze of a pet when you enter a room with its food bowl is similar to the behavior of begging lemurs (arboreal, monkeylike mammals in Madagascar) who open their eyes and stare widely at an animal from whom a desired goal is sought (Chevalier-Skolnikoff, 1973). There are also animals that do not respond at all to the stare, but this is not a natural occurrence. It is a disturbing sign of the deadening effects of captivity that zoo animals no longer exhibit aggressive responses to staring humans. "Nowhere in a zoo can a stranger encounter the look of an animal. At the most, the animal's gaze flickers and passes one. They look sideways. They look blindly beyond. They scan mechanically. They have been immunized to encounter. . . ." (Berger, 1980, p. 26).

Other significant aspects of looking behavior among primates are related to establishing and maintaining power hierarchies among the species. A common feature in the rank-ordered behavior of sub-human primates is the persistent attention by subordinate members toward the dominant member(s) of the group (Chance, 1967). Such attention may assist primates in spacing the members of a rank order in ways that maximize the cohesive organization of the group and may serve as a basic mechanism in social relations by facilitating less aggressive interaction (Exline, Ellyson, & Long, 1975). In other words, if the power relations among primates can be determined and maintained by who can look at whom, then the need for actual aggressive acts is greatly lessened.

As with animals, the human stare helps to determine dominant and submissive roles. A major function of eye contact is to establish this power relationship: The dominant person stares (i.e., engages in prolonged, direct gazing at the other's eyes or face) while the submissive person lowers the head or averts the eyes as a gesture of appeasement (Etkin, 1967; Giessler, 1913; Scheutz, 1948). There is considerable support for the relationship between direct gaze and the struggle for dominance (Argyle & Dean, 1965; Fromme & Beam, 1974; Strongman & Champness, 1968). Argyle and Dean (1965) speculated that if A wishes to dominate B, s/he stares at B with the appropriate expression. B accepts this signal with a submissive expression or by averting her or his eyes. The alternative is for B to challenge the other by trying to outstare A. The result may be similar to this familiar contest between the eyes:

> I remember one evening I spent in a bar when I was a young soldier. I happened to catch the eye of a marine across the room. We had both had a few drinks, and the idle glance suddenly turned into a contest. Who was going to outstare the other? After a short, angry locking of eyes, I told myself: this is silly unless I want a fight. I broke eye contact first, the marine took it as an appeasement signal and, with a cheerful grin, came over to my table and clasped my hand. (Fast, 1978, p. 24)

Although eye contact is usually perceived as positive and affiliative, the stare is generally indicative of aggression or hostile intent. The person who is stared at is affected both inside and out. Internally, staring results in increased physiological arousal (Gale et al., 1972, 1975; Nichols & Champness, 1971). In terms of overt behavior, staring usually results in attempts to avoid or escape from the situation. Staring at another person is an act with interpersonal implications which cannot be ignored. The stare demands a response, and in a situation where there is no appropriate way to respond, the person becomes tense and anxious to escape (Ellsworth et al., 1972).

Sometimes the stare is challenged by another and two pairs of eyes meet in a silent struggle, until one becomes dominant: "He looked into the . . . man's eyes and held him with his look, hypnotized him, made him mute, conquered his will, commanded him silently to do as he wished" (Hesse, 1957, p. 25). Some researchers argue that dominant eye behavior is an attribute of a dominant personality (such as suggested by the preceding quotation). Positive relationships have been found between dominance and "winning" in a staring contest (Strongman & Champness, 1968), and between descriptive measures of dominance and total gazing (Levine, 1972). The literature of assertiveness training reflects a belief that dominant persons look more at others than submissive persons do (Fehr & Exline, in press). Indeed, it has been found that assertive persons who "role-played" assertiveness gazed more at a confederate than the less assertive persons did (Galassi et al., 1976). Leaders have been found to sit in positions which allow easy eye contact with as many group members as possible (Greenberg, 1976). Strongman & Champness (1968) suggested the existence of hierarchies of eye gaze dominance (that is, people may be arranged in a hierarchy according to the likelihood that they would be the first to break eye contact in dyadic interaction).

Much of the psychological research categorizes experimental subjects as "dominant" or "submissive" personalities, as though a person with "dominant" personality traits would display "dominant" behavior in most if not all situations. From this perspective, hierarchies of social power would be a natural consequence of differences in the dominant and submissive aspects of individual personalities. It is, however, a mistake to assume that hierarchy arises solely from personality differences. Unfortunately, psychology in general emphasized individual traits while often ignoring the highly relevant situational factors (Jones & Nisbett, 1971; Rutter & O'Brien, 1980). While it is apparent that visual behavior is systematically affected by a dominant or submissive position in the power hierarchy (Exline et al., 1975), no one can say with certainty that the position is the result of the dominant or submissive personality traits.

Researchers have also generally ignored (or denied) the broad social factors which influence power role relationships. Objective conditions such as class, sex, or race encourage dominant and submissive behavior in interpersonal relations. Dominating behavior may be an attempt to maintain the status quo (i.e., to keep the power hierarchy in one's favor), while submissive behavior may be the only way to survive, to stay on a job, or to "keep the peace." The "proper" eye behavior for the respective roles has become almost ritualized in the contexts of some kinds of power relationships: Men openly staring at women in the street, or dropping the eyes as a sign of respect for one's "superiors"—parents, teachers, employers, judges, older adults, etc. These social dictates of nonverbal behavior reinforce the nature of the proscribed relationship. This direction of a subordinate's eye behavior is one means a superior has to exercise control.

As a general rule, the superior can tell the subordinate when and where to look. The sergeant may walk around and glare at the recruit, but the recruit must stand at attention with eyes straight ahead. An angry parent may order a child to "look at me"; or, a parent may say, "don't look at me that way." Teachers routinely use eye contact as a means of controlling the classroom. Preventive discipline may be largely unnoticed when, for instance, a teacher stares directly at a talkative student while continuing the lesson, thus sending a nonverbal message to be quiet (Grosshans, 1978).

When a child is reprimanded by a parent or teacher, s/he will commonly signal appeasement by downcast eyes. In a classroom situation, this signal is also important in relations with other students. The chances of being harassed or accosted are greatly reduced by the nonbelligerent act of looking down. A white high school student in Chicago wrote an essay for the school newspaper on how he survived while walking through a black ghetto area. To avoid being challenged by territorial gangs, he learned never to stare another adolescent male in the eyes, and preferred to keep his eyes on the ground (Freedman, 1979). In a confrontation, the one with the most control has the most "eye room," while the other—in deference, fear, or on command—has correspondingly less.

Dominant and subordinate roles are probably most easily recognized in relationships which are rigidly structured and/or based on authority, such as in school or the military, in some religious groups, and even in street gangs. Scheinfield's (1973, cited in Freedman, 1979) work with a territorial gang on Chicago's West Side showed that within the gang there was a looking order which closely correlated with the dominance hierarchy. The rule was that attention must be paid to the leadership, and to receive attention one must earn it. Thus, the leaders never looked at the lower-status males and, though the subordinates constantly watched the leaders, the lower-ranking males did not stare into the eyes of the dominant males. By noting who looks at whom, Scheinfield was able to construct a hierarchy that followed, fairly precisely, the members' ratings of one another's power within the gang. This pattern is similar to the animal behavior in social groups which was discussed earlier. Daniel Freedman (1979) found a similar trend in his experimental work with children, and has referred to this behavior as a "fearful watchfulness" on the part of the subordinate and a "disdain" on the part of the dominator.

The pattern of the subordinate looking more and the dominant looking less is a significant finding in many studies on power relationships and status (Efran, 1968; Exline, 1971; Fugita, 1974; Mehrabian, 1968a; Mehrabian & Friar, 1969). In investigating the relationship of eye contact to status relationships among college students, Efran (1968) found that first-year students conversing with a "senior-freshman" pair looked more at the senior, and Exline (1971) showed that a

ROTC student received more eye contact than a gym student. In all of these status conditions, the less powerful party pays greater visual attention to the more dominant party.

Generally, both females and males have more eye contact with a person of high status, but the difference between amount of eye contact for high-status and low-status addressees is greater for males than for females (Mehrabian, 1968a). How much one likes or dislikes a high- or low-status individual also affects the visual behavior of females and males differently. Mehrabian and Friar (1969) reported that males had more eye contact with liked high-status addressees than with disliked high-status, or liked or disliked low-status addressees. For females, however, eye contact was significantly less with disliked low-status addressees than with any of the other groups.

Status interacts significantly not only with liking but also with approval and eye contact. One might expect that the approval of the high-status party would be important to the party with less power and that the subordinate would thus spend more time monitoring the behavior of the superior for signs of approval. Although eye contact tends to be more frequent in any approval-seeking condition (Pellegrini, Hicks, & Gordon, 1970), the status of the one from whom approval is sought also appears to influence visual behavior. Gaze interaction is similar between subjects and persons who give approval and those who give disapproval, as long as the approvers and disapprovers have low status. When faced with high-status approvers and disapprovers, however, it seems we look more at the approver (Efran, 1968; Fugita, 1974).

Experiments by Exline, Ellyson, and Long (1975) have considerably increased our understanding of the variables associated with power and looking behavior. They have verified that the dominant party looks less and have identified a specific characteristic of the dominant-subordinate relationship: Power resides with the one who looks the least *while listening.* Apparently the individual in the superior position has less need to pay close attention to the speaking behavior of a subordinate, and may spend 50% less time looking while listening (Ellyson, 1974). Exline and his associates call this pattern "visual dominance behavior," which is a relatively low index of look-listen to look-speak behavior. They suggest that visual dominance behavior is a tangible element in the "command presence" of military officers and in the leadership abilities of business or government executives.

People who desire power or control over others manifest a similar pattern. An individual who wants to control another exhibits the same visual pattern of looking while speaking and listening to an equal as does a superior to a subordinate (Exline et al., 1975). Apparently this behavior reflects the expectations of our society concerning visual dominance and power, because the individual who does not look while the other person talks has indeed been judged as more powerful and more important than the individual who does look (Exline, 1971).

This judgment of power based on who looks least while listening is similar across the sexes. A recent series of studies using female dyads (Ellyson, Dovidio, & Fehr, 1981) have replicated and extended the work of Exline et al. (1975), which used only male dyads. These studies showed that subjects can differentially decode low, moderate, and high visual dominance displays. Since sex of the subject does not appear to be an important variable in either encoding or decoding visual dominance, we may generally conclude that visual dominance behavior is a valid phenomenon and is dynamically similar for both sexes (Ellyson, Dovidio, & Fehr, 1981). Implicit in all of these observations is the assumption that the "subordinate" accepts the legitimacy of the power difference. If the subordinate does not, then s/he may be neither more visually attentive to the other nor more visually submissive (to signal acceptance of a subordinate role). Exline's finding (1971) that college students look more at a ROTC student (legitimate condition) than at a gym student (illegitimate condition) seems to support the notion that someone who sees power difference as illegitimate will be visually less attentive. Generally, however, there is a set of cultural/social norms governing who may look at whom and for how long. The general feels free to eye the sergeant, the executive to eye the secretary, and the lawyer to eye the client, and it is likely that everyone would feel uncomfortable if the "who-looks-at-whom" patterns were completely reversed.

In addition to the effect of power and status, it seems reasonable to assume that certain kinds of personality traits would encourage or maintain the adoption of particular roles in a power relationship. An examination of these traits should reveal significant differences in behavior between those for whom the traits are strong or weak, and possibly indicate patterns of eye contact which are systematically related to personality differences. For example, since mutual gaze signifies at least momentarily intimate contact between two people, we might assume that individuals who are highly affiliative or highly dependent would desire more intimacy (and therefore more eye contact) than individuals who have a strong need to control others and avoid intimate relationships (Exline, 1963).

We must be careful that these personality traits are properly recognized as a matter of measurable differences rather than as a characterization of the individual. In other words, because an experimental subject receives a high score in "dominance" on a questionnaire and engages in what is defined as a "dominant" behavior, it does not mean that the subject has a "dominant personality." It is entirely possible that the high scorer on dominance will act subdued in other contexts or that a low scorer will display dominant behavior on occasion. With this caution in mind, the following section will explore a variety of personality traits that are (rather loosely) referred to as "dominant/submissive" traits.

Exline (1971) has defined these traits more strictly as "high-control-oriented" and "low-control-oriented." Although both high- and low-control-oriented subjects did not differ significantly in the amount of looking while speaking, one significant difference that he found between these types of control orientation was in response to the kinds of looks received from the confederate. The low-control-oriented subjects did not differ in the amount of visual attention they gave, regardless of how much they received. When social reinforcement, such as gaze, is reduced or withheld, it appears that the low-control-oriented individuals will search for more reinforcement from the other, while those who are highly control-oriented will respond in kind or not change their behavior at all. Low-control individuals seem to have more to fear or more information to gather from a person who does not look. Persons who are highly dependent on others exhibit the same visual pattern (Exline & Messick, 1967). High-control individuals, on the other hand, apparently respect a person who looks less. In Exline's experiment, they rated the nonlooking individual as significantly more potent than the one who looked and also rated him as more potent than did the low-control-oriented subjects. People who desire control over others seem to attach greater power to those whose eyes they cannot catch.

High- and low-control-oriented individuals also appear to interpret differentially the dominance challenge signalled by high levels of eye contact. When faced with a stare, high-control-oriented individuals may respond by approaching closely and quickly (Fromme & Beam, 1974) or by staying and competing in some way (Lawless, 1976). Less controlling people may approach more slowly or leave the situation. Low-control-oriented males, in particular, may find the high levels of eye contact a negative experience. Given the societal expectation that males will behave in an appropriately "dominant" manner, the male who does not desire that control may avoid or negatively evaluate the communication of dominance cues (Fromme & Beam, 1974).

There are, however, many situations which demand dominant or submissive cues regardless of the individual's predisposition to a particular control orientation. The sergeant is expected to keep eyes straight ahead while talking with the lieutenant, even though the sergeant may have a more "dominant personality" than the higher officer. These expectations of behavior in certain situations carry rewards for those who fulfill them. For example, a dominant personality would do well to look more, not less, while listening to an employment interviewer. High eye contact contributes significantly to enhanced credibility and favorable ratings of the interviewee (Imada & Hakel, 1977; Tankard, 1970), and interviewees are likely to be considered more sincere and more intelligent (Holstein, Goldstein, & Bem, 1971; Wheeler, Baron, Mitchell, & Ginsburg, 1979). Besides looking for the obvious job skills or expertise, an interviewer is consciously searching for signs of attentiveness and sincerity which are revealed in the interviewee's nonverbal behavior. Interviewers may be less conscious of searching for signs of submissiveness, whether the interviewee is obedient; a "team player." Of course, the one who is hired may then go on to a higher position than the interviewer; and in later encounters between them, we would expect the interviewer's eye contact to conform to a subordinate's.

Nonverbal behavior patterns reinforce the roles in socially proscribed relationships—so that it is perfectly "natural" and wholly appropriate for a man to stare at a woman who looks down, or for an executive to look into space as the employee anxiously searches her or his face. Our assumption here is that power dynamics exert more influence on eye behavior than personality characteristics.


## CONCLUSION

Historical, journalistic, and biographical accounts of charismatic persons add support to the psychologists' findings that one of the functions of eye contact is to influence the behavior of others. Whether on the campaign trail or recruiting members for a group, leaders use their eyes to manipulate interactions for their own purposes. Hypnotists play on our vulnerability to the eyes to induce suggestible states, and members of some cults essentially do the same when attempting to bring in new people.

Because of their ability to exert influence on others, the eyes are an integral part of nonverbal messages about power in our daily interactions. The language of the eyes betrays more than emotion, personality, cultural background, etc.; it reveals the distribution of power within social relationships.

There are certain eye behaviors associated with dominance and submission. In both humans and primates, when there is a struggle for dominance, the direct stare is used as a threat. Gaze aversion, in the context of a power struggle, signals submission or peaceful intent. If roles of dominance and subordination are already established (as in the classroom between student and teacher, in the military between commanding officer and enlisted person, or on the street between gang leaders and gang members, etc.) the dominant person no longer needs to control the situation with the direct stare (except, of course, when there is rebellion in the ranks). Commanding a wide field of visual space, the dominant person can look anywhere and looks less at the subordinate, while the subordinate provides greater visual attention to the dominant party, thus signalling respect, searching for approval, and monitoring the behavior of the one who has the ability to determine the outcome of the situation.

We are generally unaware of how we follow the rules of "proper" eye behavior within the social hierarchies of our lives. However, our eye contact patterns contribute toward maintaining the prevailing "pecking order," whether we are aware of it or not. Further examples detailing these patterns are provided in the next chapter, where the use of eye contact is analyzed as a mechanism of power between women and men, blacks and whites, and between other minorities and dominant majorities in United States society.

## NOTES

[1] In a number of operant verbal conditioning experiments it has been found that eye contact at the end of another's utterances makes her/him talk more (Argyle & Cook, 1976; Krasner, 1961; Reece & Whitman, 1962). Of course, eye contact will be reinforcing only if the receiver of the gaze has had mutual gaze paired with positive experiences in the past. Reinforcement is also dependent upon the history of the relationship between those engaged in mutual gaze. For example, being looked at by strangers can be very unpleasant (Argyle & Cook, 1976). We also vary in our experience of eye contact due to economic, cultural, and religious backgrounds. Some may feel that eye contact is an invasion of privacy, a sign of mistrust, or an act of defiance. Generally, however, eye contact is experienced as a signal of another's personalized attention; it provides feedback about our behavior; and it often expresses affiliative attitudes. It is probably for these reasons that eye contact functions as a reinforcer. Perhaps another reason for its power to reinforce behavior is that it fulfills "a basic need" (Hodge, 1971) to know that others are aware of our existence.

[2] Kleinke (1980) found that the effectiveness of gaze in influencing the behavior of others depended on the "legitimacy" of our requests. While airport patrons were more likely to comply with a legitimate request (e.g., for a dime to make an important phone call) when it was accompanied by gaze,

gaze carried no effect when it accompanied an illegitimate request (e.g., for a dime for a candy bar). Furthermore, in a second study, gaze actually reduced the likelihood of compliance to the illegitimate request. Kleinke (1980) proposed that to look away while making an illegitimate request may serve to make one appear more humble and thus more acceptable to a potential benefactor. Kleinke and Singer (1976) examined the effect gaze had in other situations of interpersonal influence. While handing out leaflets, two female and two male experimenters either demanded "take one"; asked in a conciliatory tone, "excuse me; would you like one?"; or said nothing at all. The experimenter either gazed continuously at the passerby to whom they offered leaflets or directed their gaze at the street. The presence of direct gaze at the subjects significantly increased the chance that the leaflet would be accepted. Unlike in Mehrabian and Williams' study (1969), gaze was particularly helpful for the male experimenter. The tone of the request had no effect when gaze was present. The least compliance occurred in subjects who received no verbalization and no gaze from the experimenter.

[3]In a student-teacher checklist developed by a student-teacher supervisor, O. R. Grosshans (1978), eye contact was listed among such skills as "knowledge of content," "use of humor," and "class control." Caproni and associates (1977) investigated whether availability for eye contact between teachers and students would increase participation in a graduate-level seminar. The instructor's position was varied to provide four different levels of eye contact availability, ranging from high to low. Results showed that students in high eye-contact areas participated more than students in low eye-contact areas.

# CHAPTER III

# Eye Contact and Relationships of Power: Gender and Minority Populations

In the preceding chapter it was proposed that nonverbal communication aids the maintenance of a society's power structure, with eye contact therefore reflecting the hierarchical relationships within that power structure. If we are to accept the proposition that eye behavior and status are related, we should demand concrete evidence of significant differences in eye behavior between the more powerful and the less powerful members of a society. In reviewing the mass of literature on patterns of visual behavior, we find gender to be the most powerful single variable; differences between the sexes are the rule rather than the exception (Ellsworth & Ludwig, 1972). In almost all cultures, the most common "power relationship" is between men and women. Indeed, it has been argued that the relations between the sexes provide the original model of interpersonal hierarchy: male-dominant and female-submissive (Firestone, 1970). Throughout history, this dominant/submissive relationship has been institutionalized in the social, economic, and political frameworks of most societies. Awareness and investigation of the male–female hierarchy have only recently become widespread. In this chapter the systematic research on the visual behavior of females and males, between and among the sexes, is reviewed in order to increase our understanding of this power relationship and how it operates.

The historical position of black people in this country as a less

powerful minority in relationship to the more powerful white majority makes black-white variations in eye contact another specific case in
which nonverbal, power-related behaviors may be studied. The second part of this chapter examines the existing research on black and
white differences in eye contact. Unfortunately, only a small body of
research has been developed in the study of black patterns of eye
contact. Hopefully, investigation will expand to provide us with a
more detailed and accurate understanding of dissimilarities between
black and white nonverbal behavior. Such dissimilarities are not
simply due to dominant versus subordinate role behaviors. Black
people also have unique heritages which contribute to the differences
between their eye contact patterns and those of other cultural backgrounds.

The final section of this chapter explores how eye behavior helps to
reinforce the stigmatization of another minority population: the disabled or, as some are now terming them, the "differently abled." The
chapter concludes with a discussion on role barriers, removal of these
barriers, and eye contact as a reflection of social change.


## POWER AND GENDER

From the moment of birth, human beings are socialized to interpersonal behavior patterns which reinforce the male and female sex roles
and their concomitant statuses. Because they are repeatedly acted
out and because we do not consider the significance of our actions,
these behavior patterns function as a means of control. The unequal
power relations between men and women are often not recognized as
such, because what is habitual is seen as desirable or necessary, or as
plain old "human nature" (Henley & Freeman, 1975). Yet, in the arena
of nonverbal power, it is one sex who generally stares, interrupts, or
touches, while the other sex lowers the eyes, shuts up when inter-
rupted, and accepts the touch. These nonverbal behaviors are traits
which society assigns as desirable characteristics of female and male
roles, and may be properly categorized as dominant and submissive
behaviors.

Women, unlike other powerless groups, are close to the centers of
power because they are wives, mothers, lovers, secretaries, and so on.
There is a constant interaction between the sexes and therefore a
constant reinforcement of the dominant/submissive roles which most
often characterize this interaction. Women's emotional closeness with
men makes them particularly susceptible to the misuse of power.
Power signals from men can often be disguised as signs of intimacy.

The deep look or a touch on the arm are behaviors which may be just as much a means for ensuring submissive, "proper" behavior as they are for expressing closeness (Henley, 1973–1974). These "little body movements, gestures, glances and facial expressions . . . contain the language of dominance and submission which maintains the status quo of male supremacy" (Henley, 1978, p. 34).

Since much communication is both about power and carried on nonverbally, an examination of the eye contact between men and women can provide us with explicit information about how our largely unconscious nonverbal behavior helps to maintain the power structure. There are two primary differences in female and male eye contact behavior: women use more submissive gestures with their eyes than do men; and women use more eye contact than do men. This pattern is found throughout much of the animal world, where low-status animals must stay attentive to the movements of the more powerful members of their species. It is also found in the eye contact patterns between human leaders and followers, subordinates, and superiors (Exline, 1971; Henley, 1977).

It has been well established that females look more in an interaction and engage in more eye contact than do males, regardless of the sex of the person with whom they are interacting (Aiello, 1972a, 1972b; Allen & Guy, 1974; Dabbs, Evans, Hopper, & Purvis, 1980; Exline, 1963; Exline & Winters, 1965a; Haviland & Malatesta, 1981; LaFrance & Carmen, 1980; Levine & Sutton-Smith, 1973). The female pattern of greater overall looking includes both more looking while listening (Exline & Winters, 1965a) and while speaking (Exline, Gray, & Schuette, 1965; Levine, 1972). This difference between the sexes has been reported at as early an age as one day old (Hittelman, Dickes, O'Donohue, & Bloomfield, 1980). Research indicates a consistent trend for both younger and older females to engage in more gazing, show a higher frequency and longer duration of eye contact, and spend a higher percentage of time in eye contact (Haviland & Malatesta, 1981). Although women look more, they usually modify their looking behavior with submissive gestures, such as looking down or frequently breaking eye contact. When men look, it is with a more direct, unbroken gaze. These behaviors are stereotypical sex-role characteristics: Staring is considered to be a masculine form of behavior and dropping the eyes when stared at is feminine.

Haviland and Malatesta (1981) pointed out the "remarkable" uniformity of the findings on sex differences in gaze behavior and the early age at which these differences become apparent. Their recent review of the research indicates that it is not clear whether the differences between females and males reflect initial motivation or absolute ability. They conclude with a number of questions about the development of sex differences in nonverbal behavior: "Why are these patterns maintained over time? Initial constitutional differences are likely to tell only part of the story. The question that one really ought to ask is what do people do when girls gaze attentively? What do people do when boys look away? Do parents interpret and respond to attentiveness or inattentiveness of male and female infants in the same fashion?" (pp. 197–198).

There are almost as many explanations for sex differences in nonverbal behavior as there are researchers studying them. To some it is a matter of constitutional differences; others focus on perceptual variables or personality variables, while still others swear by socialization and the social context in which visual behavior occurs. Women's greater amount of eye contact has usually been explained as stemming from women's higher affiliative pattern as well as their greater emotional receptivity (Exline, 1963; Henley & Freeman, 1975; Kleinke, Bustos, Meeker, & Staneski, 1973). It has been found consistently that eye contact occurs in greater amounts in people who are more affection- and inclusion-oriented (Exline et al., 1965) and women are more affiliative than men. It has further been shown that not only are females more sensitive and responsive to nonverbal cues than males (Argyle, Salter, Nicholson, Williams, & Burgess, 1979; Haviland & Malatesta, 1981; Thayer & Schiff, 1977; Zuckerman, Hall, DeFrank, & Rosenthal, 1976), they also appear to require more visual feedback than do males (Argyle, Lalljee, & Cook, 1968).

It may well be that the importance of visual phenomena in their social field is interpreted differently by men and women. For example, Witkin's experiments showed that women are more visually dependent on the physical field, both in establishing their bodily orientation in space (1949) and in locating embedded figures (1950). Along these lines, Exline (1971) has suggested the following two points: that visual information plays a larger role in the social field of women than of men; and that women's visual activity is more sensitive to social field situations than that of men.

Many researchers have interpreted women's higher gaze attention in sex-stereotypic terms, such as the expression of affiliation, field dependence, or a greater concern with others (Exline, 1963; Harper et al., 1978). Yet, when we observe visual behavior within its social context, other explanations are suggested. Nancy Henley (1977) pointed out that visual information may be more important to women because it is less likely to be available to them. Women are often excluded from informative interactions with men, and men's greater concealment of their feelings necessitates greater attentiveness. Henley also noted that eye contact is more characteristic of those who must seek approval from others, and suggested that women are dependent on men for cues on the appropriateness of their behavior. The greater visual attentiveness of women, in conjunction with their dependent role, may even be a mask for wariness. When told to conceal their feelings, women look more, while men look less (Exline et al., 1965). Perhaps women have more reason to be "on guard" when concealing or denying their true feelings. Finally, there is at least one simple reason why women look more when talking with men: The listener tends to look more than the speaker, and men tend to talk more than women (Henley, 1977).

The greater nonverbal sensitivity of females is consistent with sex role stereotypes (Broverman, Broverman, Clarkson, Rosenkrantz, & Vogel, 1970), which create expectations from women to relate to others in an expressive manner and by tending to their needs. "The oppressed status of women might make sensitivity to nonverbal communication particularly important as it allows them to read better the wishes of more powerful others" (Hall, 1978, p. 854). Indeed, the findings that women both look more and break away gaze more illustrate two aspects of the female sex-role: to be attentive to others (and alert to the male's behavior) and to be submissive. Many studies have found that women of varying ages are more accurate than men in both encoding and decoding nonverbal cues (Fujita, Harper, & Weins, 1980; Hall, 1978; Isenhart, 1978). Curiously, women may sometimes inhibit their superior decoding abilities (even though it may be to their disadvantage) because they are socialized to be more accommodating to others (Rosenthal & DePaulo, 1979; Weitz, 1976). Rosenthal and DePaulo (1979) showed that women are more guarded in reading nonverbal cues that senders may be trying to hide. Men, on the other hand, show no such accommodation.

Differences in the visual behavior of men and women may be seen by closely observing the constant interaction of the sexes. Researchers have identified gender differences in both positive and negative interactions. In conversation with someone they like, females tend to increase their looking while talking and males tend to increase their looking while listening (Argyle & Ingham, 1972). Thus, it appears that the sexes have different ways of signaling affection or intimacy. Over a period of time in a positive interaction, males tend to decrease their eye contact and females to increase it (Exline & Winters, 1965a). During a negative interaction, however, as when one is embarrassed, both genders tend to decrease their eye contact (Exline et al., 1965). Eye contact also decreases when in conversation with someone you do not like, but tends to increase with growing dislike. Males show a much greater amount of eye contact with extremely disliked males than with extremely disliked females, which may be the result of heightened vigilance (Mehrabian, 1968b).

Both females and males tend to look at their own sex more than at the opposite sex, at least in pairs. Argyle and Ingham (1972) found that male-female pairs had less eye contact than all-male or all-female pairs and reported that women show a slightly greater inhibition in cross-sex pairs. Patterson (1973) also reported a tendency for greater eye contact in same-sex pairs, while other researchers have specifically noted that female pairs spend more time in mutual gaze and in monitoring each others' faces than do male pairs (Kendon & Cook, 1969). It appears that the mutuality and reciprocity expressed through eye contact is generally shared most between two women and least between a man and woman. However, the observation that women look more may be true for pairs only and only for two people engaged in a "focused" interaction (e.g., a conversation or mutual task). Coutts and Schneider (1975) wondered whether an "unfocused" inter-action, in which subjects sat apart and were forbidden to talk, would influence the findings on gender differences. Contrary to the research using "focused" interaction, they did not find that females looked more in general, although female dyads engaged in the most glancing. Schneider, Coutts, and Garrett (1977) extended this questioning to determine whether interacting with two other people, rather than one, would similarly influence the results. Again, females did not gaze more than males and neither gender seemed visually to favor same-sex others. This evidence suggests that group size may moderate the influence of gender on visual interaction patterns.

Other situational variables seem to have a pronounced effect on the visual behavior of the sexes. Exline (1963) showed that in a competitive situation, females with high affiliation needs reduced their eye contact much more than highly affiliative males, while females with low affiliation needs increased their eye contact much more than males with similar needs. In a situation which involves a certain competition for space (i.e., crowding) eye contact decreases in both sexes but even more when crowded by a male (Dabbs, Johns, & Powell, 1976). It has been shown that crowded females tend to become more friendly and positive towards one another, while crowded males become negative and unfriendly (Freedman, 1975). Perhaps this finding is related to females' greater ease and comfort with eye contact at close physical proximity. Females will look more and look longer than males at close distances, but the reverse is true at farther distances (Aiello, 1972a). In any case, it appears everyone would prefer to be crowded with females or, at least, would evaluate the experience more favorably than being crowded with males (Dabbs et al., 1976).

Gender differences are apparent in the dynamics of eye contact and violation of personal space. Buchanan, Goldman, and Juhnke (1977) used a public elevator to examine the relationships among varying degrees of eye contact, sex of participants, and violation of personal space. In one experiment, naive subjects were forced to violate the personal space of two men or two women confederates who were either looking or not looking at the subject entering the elevator. Males violated the space of the confederate who averted gaze, regardless of gender, while females chose to violate the personal space of the female who was looking and the male who was not looking as she entered. Males also showed no preference in violating the space of male and female confederates who directed their gaze at incoming subjects but, again, female subjects chose to violate the space of the female who looked.

The results of the elevator experiments suggest that women find a man's look to be threatening and are therefore more comfortable in approaching a woman who is looking. Men, however, seem less sensitive to the gender of the other, preferring only a nongazing to a gazing other. Perhaps eye contact is less important to males, outside of a staring confrontation, in signaling a threat. Apparently the nonverbal behavior cues which indicate dominance occur more often in men than in women (Frieze & Ramsey, 1976) and it may be, too, that men and women use different types of proxemic behaviors to signal their position in a dominance hierarchy. Fromme and Beam (1974) showed that men with a high need for dominance make little use of eye contact as a signaling device as they approach others, but rather, communicate their dominance through personal space and rate of approach (i.e., a closer and quicker approach). However, women who are highly dominant tend to meet a direct gaze with a direct gaze and also tend to avoid approaching too closely or too quickly. One study suggests that the need for dominance has more effect on gender differences in eye contact than the need for affiliation (Gray, 1971). Given this premise, dominant women would have the greatest amount of eye contact and dominant men would have the least amount. However, another study suggests that *both* men and women have high needs to control others (Exline et al., 1965).

Struggles for dominance sometimes occur when partners are unable to see one another—consider business executives wheeling and dealing by telephone. Visibility effects any communication and it appears that a lack of visibility particularly effects interactions between the sexes. One study on visibility and gender reported that subjects had more total pause time and more long pauses in speech when conversing with an opposite-sex experimenter who was hidden from them. They also had fewer body movements and less liking for the experimenter. Subjects with same-sex experimenters were not affected by the lack of visibility (Wolf, 1976).

Studies have reported that women become uncomfortable if they are unable to see the person with whom they are conversing and would prefer to see even when they themselves were concealed (Argyle, Lalljee, & Cook, 1968; Argyle & Williams, 1969). This finding complements another, that women look *more* when they are concealed than when they face another person (Argyle & Ingham, 1972). It may be that women's discomfort results from their greater reliance on visual feedback; without such feedback they may find it more difficult to adjust their behavior and respond "appropriately" to the other (Ellsworth & Ludwig, 1972). Or it may be that the lack of visibility inhibits women's greater tendency for affiliation. Although women usually prefer to exchange more personal information with strangers than men do, this preference occurs only when the other person is visible (Exline, 1962). In cases where the other person is visible and the subject is not, males will increase and females will decrease their amount of speech by as much as 40% (Argyle, Lalljee, & Cook, 1968). Men talk more when concealed *and* paired with a woman. Women, especially in talking with men, feel more observed than men. Women in one of the visibility studies reported that they were the "performers" and men, the "audience" (Argyle et al., 1968).

Like an audience which is permitted to stare unblushingly at the actors on stage, so too is the male permitted to direct his stare at the female, usually to her resultant discomfort. "When confronted with the direct gaze of a male . . . I did experience those physiological 'alarm signals'. I felt that the male stare seemed 'invading,' . . . like an unwanted intimacy," was how one researcher described her experience during field experiments on staring (Freedman, 1979, p. 78). The fact that women are more frequently stared at than men are, and by men, is seldom questioned by either sex. The practice is noted in our language, with specific words for men staring at women, such as leering or ogling (Henley, 1977).

Women, of course, stare at men as well, but will often do so in a subtle if not submissive way by slightly tilting their heads or by interrupting their looking behavior at men by furtive glances away. In another field experiment on staring (Smith, Sanford, & Goldman, 1977), female confederates were required to stare at fellow students for fifteen minutes, thereby behaving in a visually nonstereotypical manner. Female students who were stared at quickly departed the area, particularly when the starer was one of the male confederates. Male students, however, returned the stare more frequently, especially when stared at by a female. The female confederates reported that they felt their staring was a "novelty" to men. The novel aspect of the female stare was taken more as a challenge than as a threat by the male students. Many of them tried to break the female's fixed expression by smiling, speaking, or other overtures.

As a dominance behavior, the stare may act as a social gender signal and serve to proscribe the nature of the interaction between the sexes. It is socially accepted, and thus *expected*, that men will stare and that women will avert their eyes before the man's gaze. Other expectations about socially-proscribed gaze patterns may also influence behavior toward the opposite sex. When people behave as we expect, we tend to like them more and find them more attractive. How attractive we find someone of the opposite sex may depend, at least initially, on how well he or she conforms to expected gaze patterns.

The social gender signals of low gaze by men and high gaze by women, and the expectation of these signals, probably influences everyone's initial impression of a person of the opposite sex. Attractiveness, sincerity, and attentiveness are among the host of characteristics which may be conveyed by an expected gaze pattern. Indeed, we may assign these characteristics to others not only when their gaze conforms to the pattern but also when *our* gaze toward them conforms. In one study on gaze (Kleinke, Bustos, Meeker, & Staneski, 1973), both female and male subjects found their opposite-sex partners more attractive when they thought their partners displayed sex-typical gaze patterns *and* when they thought their own eye behavior did.

In Kleinke et al.'s experiment (1973), subjects were told either that their partner's gaze had been measured during interaction and was higher, lower, or average compared to most people, or that their own gaze toward their partner was higher, lower, or average. Females and males reacted in exactly the opposite manner to gaze attributed to themselves and to their partners, as the stereotypical pattern would lead us to expect. Males who thought that their female partners had looked more rated them as more attractive, and females who thought that their male partners had gazed less rated them as more attractive. Males who were told they looked less at their partners rated them more favorably and females who were told they looked more rated their partners as more favorable.

There is, however, a significant minority of the population for whom social gender signals, such as gaze patterns, may have a different impact. All of these studies with opposite-sex partners assume, of course, that the subjects are "normal," average people, and therefore heterosexual. However, normal, average people who are gay have received absolutely no attention from the majority of researchers. It is interesting to speculate that because gays have long been forced to conceal their affiliational or sexual preference, gay men and lesbians may have developed a greater sensitivity to nonverbal cues than heterosexual men and women. Denied the intimacy of touching each often with their eyes (Webbink, 1981). As Joan Nestle described: "A more hidden people, we used eye searching to find each other; we dared to look longer than we should and, through this brave journeying of our eyes, we found one another. It was a secret language that gave permission for a secret act" (Biren, 1979, Foreword).

We may also speculate that the dominance signals emitted by men and the submissive signals of women may be largely absent from the interaction of same-sex pairs. Because gay couples receive no conscious social reinforcement for one partner to dominate or the other to submit, nonverbal signals of dominance and submission may have less meaning. It is unfortunate there has been so little exploration of these issues. If much of human society is predicated on a hierarchy of relations, such as sex roles, and there is an identifiable group for whom this pattern may not be the same, then it is important to examine in detail the nature of these apparently more equal relationships; including, of course, the nonverbal language underlying, amplifying, and clarifying overt behavior (Webbink, 1981).

It would be of further interest to examine the eye contact patterns between heterosexuals and admittedly open gay males and lesbians. Would a stereotypical negative bias influence a heterosexual's amount of eye contact with a gay person? One study (Zeiger, 1978) found that women who have negative attitudes toward lesbians will decrease their eye contact with another woman if they believe her to be lesbian. Too little consideration has been given to the role of eye contact and other types of nonverbal behavior in the interactions of different power group members (particularly in terms of stereotypical interpersonal judgments).

Our socialization to sex-typed behavior leads us to expect and desire nonverbal behavior congruent with the gender stereotypes. If our society consistently rewards men who stare or women who avert their eyes, then we may expect less tolerance toward people who do not display these patterns. For example, when a woman does *not* exhibit submissive mannerisms—when she does not tilt her head or smile deprecatingly, when she returns stare for stare—her behavior may be taken as a sexual advance, or her assertion of power may otherwise be denied by men (Henley, 1976). There are social consequences as well. Women who use nonverbal behaviors considered inappropriate to their sex are negatively evaluated by both men and women (Mehrabian, 1972). Thus, women who use a stronger power signal may very well incur rejection (Frieze, 1978).

Men who do not fit the masculine stereotype may have subjective experiences similar to women. Low-dominant males have negatively rated high levels of gaze from others (Fromme & Beam, 1974), and dependent and insecure men have reported feeling more observed when they are with females (Argyle & Williams, 1969). Men may even be disliked for exhibiting nonconforming sex-role behavior such as expressive nonverbal cues. In an experimental classroom situation (Esp, cited in Horn, 1978), students liked the male teacher who did not smile or look directly at them more than the male teacher who did smile and look directly. Liking or disliking the male teacher did not affect learning since the students learned equally well from the male teachers, whether they acted "positive" (looked directly, nodded, and smiled) or "negative" (did not smile, look, or nod). Interestingly, students learned the most from a positive woman teacher and the least from a negative woman teacher. (We may wonder: Do students learn best from teachers who act like mom and dad act?)

Reversals in the power hierarchy of the sexes can cause contention between men and women. In studying the ecology of twelve Chicago restaurants, Whyte (1949) discovered that conflict was common between female waitresses and male cooks. He attributed part of the problem to the power imbalance caused by the waitresses handing orders to the more highly skilled and paid cooks. The conflict was swiftly resolved when direct face-to-face contact between cooks and waitresses was eliminated. Usually, of course, the tension between female/male or subordinate/superior interaction cannot be eliminated so easily, especially in cases where a female subordinate has legitimate grievances about the behavior or practices of a male superior. For example, a male factory owner might glare at a female factory worker who is complaining about safety conditions in the workplace. His glare is an effort to demonstrate power and the feeling it conveys could be disdain, pride, anger, distrust, self-confidence, or sexual or class contempt. The worker might respond with emotional messages arising from her subordinate economic and social position, such as fear, hostility, or frustration. If, however, social conditions change so as to shift toward increasing power for the workers—in a strike, for example—then the female worker might be in the position to "stare down" the boss with feelings of defiance and anger.

Often a woman's economically dependent and therefore subordinate status (as wife, secretary, assistant) will keep her from even asserting herself. But what happens with the women whose status is not subordinate; that is (though they are relatively few), the employers, judges, and mayors who are women? We simply do not know what difference this change in status makes in the nonverbal behavior of women in high-status positions. Would her nonverbal behavior reflect her status or sex, or both? Do men who complain about the "bitch boss" feel threatened by the woman's position or is it an interpersonal dynamic as well? It was noted earlier that female dyads can encode "visual dominance behavior" (i.e., looking less while listening) as well as male dyads (Ellyson, Dovidio, & Fehr, 1981). But it is purely speculation as to how a man responds to a woman who is (nonstereotypically) visually dominant. Perhaps men who are stereotypically masculine will escalate their own visual dominance to counteract or challenge the "threat" of a woman's display. Or it may be that women's visual display is subtle enough to escape such a challenge.

An alternative to the suggestion that women's visual dominance may be sufficiently subtle to escape challenge is the possibility that women's dominance displays may go unchallenged simply because they are unrecognized. It is not a matter of subtlety but of sex-role stereotypes; a woman is not expected to display visual dominance, so when she does not no one notices. In a situation of visual dominance, such as a leadership position, a woman who is displaying visual behavior appropriate for leadership may still not be recognized as a leader. An unconscious perceptual bias may be operating, based on the stereotype that "women aren't leaders," which prevents us from seeing the nonverbal leadership cues displayed by women (Porter & Geis, 1981). The woman with leadership skills who wants to occupy such a position is faced with an unfortunate dilemma: if she displays the moderate demand signals which are effective for men, she may be ignored. However, if she increases the strength of her signals, she is likely to be considered overly emotional, strident, or abrasive.

The traditional gender roles of femininity and masculinity are not polar opposites; in fact, many men and women display behavior of the opposite sex when it is appropriate to the situation (e.g., women as leaders and men as nurturers). These individuals may be considered androgynous, as opposed to the sex-typed roles of feminine and masculine. "Androgyny" as a psychological construct has generated much controversy (Crosby, Jose, & Wong-McCarthy, 1981; Ickes, 1981; Lott, 1981; Pedhazur & Tetenbaum, 1979), but here the term is being used informally. If we define androgyny as an orientation toward social interaction, one combining the masculine instrumental and feminine expressive orientations (Kaplan & Sedney, 1980), we can examine how one's orientation influences visual behavior.

Ickes, Schermer, and Steeno (1979) found that, in same-sex dyads, androgynous pairs spent a significantly greater frequency and greater total duration in both eye contact and looking than did the sex-typed pairs or the pairs of one androgynous and one stereotypic person. Studies have shown that in unstructured interactions, dyads composed of androgynous individuals consistently display more involvement with each other than do sex-typed pairs (Ickes & Barnes, 1978; Ickes et al., 1979). Furthermore, androgynous pairs rate their interactions as significantly more positive and satisfying (Ickes et al., 1979; Lamke & Bell, 1981). Masculine men and feminine women seem to have the least eye contact and the least interactional involvement (Ickes & Barnes, 1978). This finding would argue that compatibility between the sexes may result from divergence from the traditional roles, rather than adherence to them.

It is possible that as society becomes more tolerant of divergences from typical sex roles, more and more androgynous behavior will appear. That is, as the respective roles of men and women change, we can expect to see more men who are comfortable exchanging a great deal of eye contact with others and more women who are comfortable using a direct, assertive gaze or stare. Men and women who value androgynous behavior (or, alternatively, those who devalue sex-typical behavior) may develop significantly different nonverbal patterns than men and women who value traditional sex roles. It is also possible that, as more and more women enter positions of power and as more and more men learn to accept women in these positions, the visual behavior of both will change. "The gender effect should diminish if our society is able gradually to shed its ingrained gender role stereotypes and patterns of male dominance" (Hall, 1978, p. 854). But there is a real concern here: Can men and women consciously change their largely unconscious nonverbal behavior to reflect their new visions of themselves as nondominant or nonsubmissive? Or will nonverbal sex differences continue to perpetrate sex-role stereotypes because people are unaware that their mouths might be talking about acting in new roles but the rest of their bodies are acting in the old ones?

## BLACKS AND WHITES

Certainly the nonverbal behavioral differences between citizens of the United States who are black (the less powerful minority) and those who are white (the more powerful majority) would be a fruitful area of study. However, only a small amount of work has been done on the subject, and extensive investigation is necessary for specific descriptions of black/white differences and interactions. The limited body of research we do have yields both provocative and contradictory results. It can be tentatively stated that blacks exhibit a tendency to engage in less overall eye contact than whites (Exline, Jones, & Maciorowski, 1977; La France & Mayo, 1976; Marcelle, 1977). Although blacks engage in less eye contact overall, they gaze a greater number of times than whites. More frequent eye contact suggests that the length of gaze is shorter and that contact is broken more often. In particular, black females gaze at both blacks and whites significantly less than white females, although more than both black and white males (Exline et al., 1977; Fehr & Exline, 1978; La France & Mayo, 1976).

In a study by Mayo and La France (1973), films of a naive black graduate student conversing for 10-minute intervals with a white or a black partner revealed a complete reversal of common white speaker-listener eye behaviors in the black–black conversational dyad. Both black speakers tended to look more while talking rather than while listening.

When blacks and whites interact, therefore, these differences may give rise to communicational breakdowns. Because the black persons looked while talking and the white persons while listening, the mutual gazes during the interracial conversation were significantly longer than for the black–black interaction.

In 1976, La France and Mayo observed the gaze behavior of blacks, whites, mixed races, and mixed-sex dyads conversing in cafeterias, hospitals, and airports. Dyads were coded for looking while listening. All-black dyads looked less while listening than all-white dyads, especially black male pairs. However, black female–male dyads did not look significantly less than white female–male pairs. Although Exline et al. (1977) did not find that blacks gaze less at their partners while listening than while speaking to them, they did report that black men showed a greater willingness to engage in eye contact at near rather than far distances, especially when the topic of conversation was intimate rather than superficial. White men tend to do just the opposite—they have less eye contact when physically closer or conversing intimately with another person.

Interracial interactions have been studied more extensively than black–black interactions. Descriptions of black relationships can assist us in understanding the nonverbal behaviors common in black culture. Kenneth Johnson (1971) described two types of behavior, one of which is more prevalent among black females and the other among black males. Both behaviors are responses, albeit of very different natures, to the power or authority of another person.

Johnson (1971) noted that in conflict situations, especially when one is in a subordinate position, black people can express a hostile, insolent disapproval of the person in authority by "rolling the eyes." This quick movement of the eyes is used particularly by black females. Perhaps rolling the eyes is partly responsible for the saying used by many blacks: "Don't look at me in that tone of voice." Another eye behavior, used particularly by black males, is avoiding eye contact with a person in an authority role. Johnson pointed out that, in black culture, this avoidance is a way of communicating recognition of the dominant–subordinate relationships. For example, many black children are taught not to be disrespectful by looking an older person in the eye when he or she is addressing them. This pattern of eye avoidance as a matter of respect is common in other cultures, such as in Japan and western Africa.

Eye avoidance may also be learned early by American blacks for self-protection in white society. According to the "rules" stemming from slavery, a black person was not permitted to look at a white person except when first looked at by the white, and then only for a short period of time. This arrangement gave the white person superiority in perceiving the visual field, including the black's behavior, while denying the power of close observation to the black person. It may also reduce the threatening aspect of a black's direct gaze. The black person's dark brown iris and pupil in the midst of the white sclera stand out sharply, which make the eyes more striking and eye movements more apparent.

H. Rap Brown described his brother's arrest for "staring at a white girl in a house down the street; he had gone down there to see a friend of ours and some white folks called the police on him" (1969, p. 58). In the South, a black man's look at a white woman has sometimes been treated as rape, as in the case of Emmett Till, the 15-year-old black youth who was lynched in Mississippi in the 1950s for staring at a white woman. Perhaps one of the principal reasons blacks may not have made eye contact with whites was their fear of reprisal, as demonstrated by these extreme examples. Another motivation might be to avoid seeing hatred in the eyes of a white person, such as the "hate stare" identified by Erving Goffman (1963), a frank expression of hostility from Southern white to black.

Other factors may also influence the gaze behavior of less powerful groups. Like the situation with men and women, it is questionable whether the more limited use of eye contact by blacks is attributable to race, culture, status, or a combination of factors. Blacks in general have a lower socioeconomic status than whites and this factor may strongly influence their looking behavior. For example, a study on the interaction between 5-year-old children and their mothers showed that high socioeconomic pairs had more eye contact than low socioeconomic pairs (Schmidt & Hore, 1970).

The differences described here in the visual patterns of blacks and whites are probably indicative of subcultural code differences in conversational behavior. Charles Payne (1977) noted that as a black graduate student, there were occasions when not engaging in eye contact caused doubt about his interest in the conversation. However, according to the dictates of his culture, his eye behavior in conversation was wholly appropriate. "In my culture to show real interest in what someone was telling you, instead of looking the person straight in the eyes you would find some object in the room to fix your eyes on and hold your head down slightly, but both the head and eyes are held very still. To look straight at the individual was considered rude, or you might be accused of staring. . . ." (p. 42). Johnson (1971) suggested that the nonverbal communication patterns common in black, but not white, culture possibly have their origins in African patterns or were the result of the black population's isolation from white society. Whatever the causes, these differences produce problems in interracial encounters, both in their own right and in the interpretations made of the differences.

If whites look more while listening and blacks look more while speaking, many nonverbal cues are going to be missed. The simple aversion of gaze by blacks is sometimes interpreted by a white person as an insult or a token of inattention. Since for whites looking away implies cognitive difficulty (Exline, Ellyson, & Long, 1975), the black person's eye avoidance may be interpreted by the white person as an inability to comprehend. This interpretation may have dire consequences in academic or employment situations when, as is most often the case, a white person is in the position of hiring or teaching a black person.

Given these subcultural differences in eye contact, it is not surprising that sometimes whites feel that blacks are avoiding their eyes, while blacks meeting with whites sometimes feel stared at (Erickson, 1976). Interpersonally this difficulty may inhibit, for example, the growth of a friendship between a black and a white woman. Since white women indicate their liking by gazing more and interpret another's gaze as meaning liking, they may possibly misinterpret the lower level of gaze by the black woman as an indication of a negative evaluation (Fehr & Exline, 1978). Unfortunately, the relationship might never progress beyond this initial misunderstanding.

Although differences in nonverbal behavior may not *cause* prejudice, they do not facilitate interracial relationships. Davis' observations (1971) on a film made at a nursery school poignantly illustrate some of the consequences of a miscommunication between the races:

> During a ten-minute sequence, one little black girl tried, by actual count, thirty-five times to catch the white teacher's eye and succeeded only four times, while a white child made contact eight times out of only fourteen tries. And it wasn't a case of favoritism. Analysis showed that the white child's timing was simply better: the little black girl kept looking at the teacher even when she was engrossed in helping another child, while the white girl saved her bids for attention, for the most part, for times when the teacher was free. Again and again in this film the teacher tried reaching out to each of the two little black girls but each try at contact turned into a near miss, either because the woman herself hesitated, as if not sure her touch was welcome, or because the child, with a slight, graceful ripple of the shoulders, shrugged off her hand. (p. 118)

Filmed samples of teacher-student interactions were also used in a systematic study of teachers' nonverbal behavior in relation to the race of their students (Feldman & Orchowsky, 1979). Naive judges rated the silent films for how pleased teachers appeared to be with their black and white students. Teachers expressed more pleasure nonverbally with white than with black students. Word, Zanna, and Cooper (1974) designed a study to test whether whites interacting with blacks would emit nonverbal behaviors corresponding to negative evaluations. They also wanted to determine if this behavior would cause blacks to reciprocate with less immediate behaviors. They found that black, as compared to white, job applicants received less immediate nonverbal behaviors (and less interview time) from interviewers in an experimental setting. In a second experiment, white interviewers were trained to emit to white job applicants the same, less immediate nonverbal behaviors that the black applicants had received in the first experiment. Subjects treated like blacks were judged to perform less adequately than subjects treated like whites. The "black" students also reciprocated with less proximate positions and rated the interviewers as less friendly.

The work of Word et al. (1974) illustrates how nonverbal behaviors can act as mediators for self-fulfilling prophecy in interracial social episodes. A vicious cycle is set in motion when a white person's racist attitudes cause her or him to emit less immediate nonverbal cues toward a black person. The black, in turn, reciprocates with less immediate cues, which the white person takes as "proof" or support for racist attitudes. Thus the cycle continues. A better understanding of the dynamics involved in nonverbal interaction between blacks and whites will not stop this cycle completely, but a greater awareness of our nonverbal behavior may help to interrupt it. Nonverbal behavior can also reveal a sense of equality or lack of racial prejudice. Penner (1971) found a significant correlation between eye contact and a positive attitude toward civil rights among white students who engaged in conversation with a black confederate.

As the power relations between American blacks and whites have shifted in the last three decades, we can speculate that so too have eye contact and nonverbal communication patterns changed in response to this shift. One might investigate this speculation by comparing the amount and type of interracial eye contact among blacks who have positive or negative evaluations of themselves in relation to whites. Regional differences in eye contact among blacks and in biracial groups may also reveal differences which reflect changes in attitudes and practices since the civil rights movement began in the 1950s.

The United States is home to a number of large minority populations, and all of them bring their own cultural code differences to interactions with members of the majority. For example, Sarah Greenfield (1979), a counselor and teacher as a Junior High School in Arizona, noticed differences in eye contact between Navajo and Anglo students. She said that most Navajo students tended to avoid "the direct, open-faced look of the Anglo" (p. 1), and that the more upset a Navajo student is, the less likely s/he is to make eye contact.

When two people from different races or cultures have difficulty in conversation, and begin to feel uncomfortable or uneasy with one another, they do not ordinarily perceive the problem as arising from subcultural differences in nonverbal codes. Unfortunately, such differences make it more likely for us to have negative evaluations of each other, and our misunderstandings help to sustain stereotypical interpersonal judgments and contribute to conflict in interracial and intercultural situations (Johnson, 1971; La France & Mayo, 1975). Increasing interracial and intercultural contact will help to reduce conflicts, as our sensitivity to (and appreciation for) our differences continues to grow. Perhaps we should have training classes in "intercultural competence," especially for those people most likely to lack such competence; that is, members of the cultural and/or racial majority (Erickson, 1976). Certainly, members of majority populations should not always expect members of minorities to assimilate the ways of the dominant culture. Whites, for example, can educate themselves about the ways of other races and learn to discard the stereotypes they have associated with those who do not behave as they do.

Judith NineCurt (1976a,b), a Puerto Rican educator, has developed a training course on cultural awareness for the use of teachers in a classroom setting. She advises that both black and Hispanic children be taught how to switch nonverbal "cultural channels," so that they can relate to and be perceived accurately by whites. "Coaching" minorities on their communication styles might be extended to other institutions besides the schools, such as the criminal justice system (Davis, 1979). Minority defendants do not usually face a judge or jury of their peers. Lawyers may tell their clients to "dress up" for trial, but never think to tell them to look the judge in the eye. Sometimes their freedom may depend on it.

## EYE CONTACT AND THE STIGMATIZATION OF MINORITIES

Power imbalances in society have resulted in the stigmatization of some groups of people, based on skin color, sexual preference, disability, and so on. Erving Goffman (1963) observed that "normals" tend to avoid stigmatized persons. In one of the few studies on homophobia (Zeiger, 1978), a heterosexual confederate talked with women subjects about school for a few minutes before opening her vest to reveal a button on her shirt which declared: "How *dare* you presume I'm heterosexual!" During the next few minutes of conversation, some of the evidently homophobic subjects shifted their bodies so they no longer faced the confederate, reduced their amount of speech, and/or refused to look at the confederate.

A group of people who are more obviously different may have just the opposite problem; it isn't that people don't look, it's that they look too much. In public places, the "differently-abled" person "will be openly stared at, thereby having their privacy invaded, while at the same time, the invasion exposes their undesirable attributes" (Goffman, 1963, p. 66). A number of studies have shown the existence of a nonverbal stigma effect. For example, able-bodied persons ended interaction earlier (Kleck, Ono, & Hastort, 1966) and moved less (Kleck, 1968) with an apparently disabled person (a mock leg amputee). Kleck (1968) expected to find a decrease in eye contact among students interacting with a disabled peer, but instead found that the students looked *longer*, at least while listening, than at an able-bodied person.

It may be that (aside from staring in the street out of curiosity) we look longer at the differently-abled person because we feel a greater need for information about this "unfamiliar" person and/or because while listening, we feel it is embarrassing to look at the person anywhere except the eyes. However, though we may look longer, apparently we do not sit more closely to a "disabled" person (Kleck et al., 1966), unless they are visually impaired (Hardee, 1976). If visual interaction is reduced or nonexistent, as with the visually impaired or blind, then perhaps we feel more comfortable about approaching. Such behavior would lend support to Langer, Fiske, Taylor, and Chamorvitz's argument (1976) that when confronted with a "physically challenged" person we may experience a conflict between a desire to collect information about a "novel" stimulus and a fear of the social sanctions against staring at others.

"Physically challenged" persons often are such "novel" stimuli for many of us because of the lack of opportunity we have to see them in the media. When exposed to images of persons different from ourselves, we have an opportunity to gather visual information at our leisure, without invading the privacy of another or giving them reason to feel like a "freak." Langer et al. (1976) asserted that staring at publicly displayed photographs is permitted by social norms. In a study in which they measured the length of visual fixation on pictures of "physically challenged," pregnant, and "normal" persons, females stared longer at the photos of the "physically challenged" persons as compared to photographs of "normal" females. Pregnant women were apparently "novel" stimuli as well, for the photos of pregnant women were stared at longer than those of "normal" women, but less than photos of those otherwise "physically challenged."

Social norms against staring at publicly displayed photographs of "physically challenged" persons seemed to operate for male viewers. When male participants were overtly observed viewing the photographs by a female confederate, they gazed more at the pictures of "normal" males than at the photographs of physically challenged males. When they were unobtrusively observed, this viewing pattern was reversed.

Despite the operation of some social norms against publicly staring at photographs of "physically challenged" persons, there had been, at least, some exposure to the "novel" stimulus. In another study, Langer et al. (1976) hypothesized that preinteraction visual access to a "physically challenged" or pregnant female would reduce stimulus novelty. Indeed, when males and females had viewed the female confederate before an interaction, they sat closer to the confederate than those who did not have preinteraction visual access.

All of these studies focus on those who are able-bodied, yet those who are physically challenged are the ones who know what it means when strangers, friends, even family and lovers react with a stare or sit farther away:

All of a sudden you become a freak or someone no longer recognizable. You're still the same person, but somehow all that is seen of you is your disability. People avert their eyes as if they are ashamed that they are still walking around. And that if they should smile at you it would be misinterpreted as mocking. Don't use a double standard in reacting to a handicapped person; we have the same feelings as anyone else. . . . (Witherow, 1981, pp. 2–3)

There has been some investigation into the differences between those who communicate in American Sign Language (ASL) and those who communicate in spoken English.[1] In order for communication in ASL to be effective, the addressee must consistently gaze at the speaker. One person cannot "say" something and be "heard" if the other is not looking (Baker, 1976). Therefore, the differing use of gaze among deaf and hearing persons is a primary problem in deaf/hearing interactions (Baker, 1976). Periods of eye contact extending longer than five seconds are not uncommon in ASL interactions. Deaf persons often feel that hearing persons are inattentive, uninterested, or being hostile because they avoid the intimacy of mutual eye gaze (Baker, 1977). Problems in synchronization can also arise in deaf/hearing interactions due to the fact that unlike hearing persons, deaf speakers signal their desire to continue their turn at clause boundaries by not looking at the addressee at all (Baker, 1976).

As more and more of the differently-abled enter mainstream society or maintain their positions after disability, and therefore become more familiar to others, we might expect that visual behavior toward them will become more like the attention given to the able-bodied.[2]

## CONCLUSION

In this chapter and the previous one, we have seen how eye contact reveals the power balance—the relationship between dominant and subordinate. Power is exercised through control and one mechanism for control is nonverbal behavior. It is the prerogative of the powerful to tell others how and where to look and not look (Argyle & Cook, 1976), and it is their prerogative to choose the mode of visual interaction, ranging from staring at the powerless to looking away and ignoring them. The powerful can treat others as if they were not there at all, a kind of "nonperson" treatment for objects not worthy of a glance (Goffman, 1963). Or they can use the power of the eyes to intimidate subordinates, such as one American employer intended when he hired a man to come into his business every so often and glare at his workforce, apparently to keep them hard at work (Davis,

1971). Subordinates have to be careful to look enough to find out the intentions of the powerful but not to look so much that it will provoke a challenge.

Both objective and subjective conditions contribute to the nature of a relationship and the particular roles played by the interactants. Researchers need to examine very closely the specific context of a given interaction to see how socially proscribed conditions affect behavior and establish more or less stable features of the relationship (i.e., who will act in which role). We also need to look at the interpersonal orientations and/or personal qualities which individuals bring with them to such a context (Exline & Fehr, 1979). Knowledge of these conditions allows us to predict the probable kinds of behavior interactants will display and the dominant or subordinate roles they will fulfill.

Since nonverbal behavior is the major factor in communication, being over four times as informative as verbal behavior (Argyle et al., 1982), it behooves us to pay close attention in a relationship to how nonverbal behavior is communicating dominant and submissive roles. Some specific behaviors have been identified, such as the dominant looking less while the subordinate speaks, or the subordinate looking away when the dominant stares. Such patterns indicate acceptance of the respective roles and serve to reinforce the roles every time they are repeated. When these roles are proscribed by society as the appropriate or "natural" behavior for certain kinds of people, they serve as a means for social control. The messages of power communicated by nonverbal behavior help to maintain the power relationship and keep the interactants in their respective "places."

Social stereotypes help to support the status quo by influencing our perceptions of people and expectations of how they should behave. Early learned stereotypes operate to bias perceptions before they register in conscious awareness (Porter & Geis, 1981), and we all believe that our perceptions are true and correct. This process is illustrated by the study mentioned earlier on women and nonverbal leadership cues. In that case, a woman who presented a nonverbal leadership cue did not receive the same leadership recognition as a man presenting the identical cue (Porter & Geis, 1981). Even the subjects who were consciously supportive of women's leadership failed to attribute leadership attributions to a woman: the feminist subjects did not view her as any more likely a leader than the male subjects. This problem identifies a major direction in which research is needed. We need, first, to identify behaviors or social conditions that will effectively dispel stereotypes or reduce discrimination, and then we need to inform people of them.

For the ones who are kept in their "place," as opposed to the ones who define them, the findings presented in these two chapters have certain implications. If the powerless want to counteract the dominance expressed over them, they must understand the mechanisms of power, including the nonverbal communication of dominance and submission. The nonverbal symbolism of power is difficult to resist unless one is conscious of it. Although knowledge of the behavior of the powerful will not change the fundamental relationships of power in our society, consciousness of them will enable the less powerful to detect some of the ways they are being controlled or coerced (Henley, 1973–1974).

It is probable that as socially prescribed roles are redefined by a society, nonverbal behavior will change to reflect the new values attached to a role. Struggles for liberation by a group of people will likely produce changes in visual behavior both among the members of the group and in their interactions with the dominant society. There is evidence that changes in power can effect changes in nonverbal behavior (Mayo & Henley, 1981). The civil rights movement in this country has reduced the incidents of hate stares by whites toward blacks and of downcast eyes by blacks. The women's movement has not only encouraged women to be more assertive, but has also encouraged men to be more emotionally expressive. Attempts to change one's behavior, as with feminist consciousness raising and assertiveness training, are attempts to embody one's own values, and not those which society has prescribed for you.

It is clear that nonverbal behavior is both a barrier to change and a facilitator of change. Mayo and Henley (1981) have outlined the evidence for both positions. The nonverbal domain can prove resistant to change because: (1) we are unaware of so much of our behavior; (2) nonverbal behavior is learned and heavily reinforced; (3) nonverbal behavior encodes power well; and (4) deviant behavior is punished. On the other hand, the nonverbal domain can serve to facilitate expressions of change because nonverbal behavior is complex (which allows certain cues to be added to or removed from a nonverbal display) and because nonverbal behavior is situationally variable (so that there is some latitude in selecting situations for innovation). The individual who seeks to change her or his nonverbal display may be most successful by bringing a given behavior to awareness, trying to alter cues related to that behavior, and doing so in a supportive (or at least not hostile) situation.

People who consciously reject many of the traditional values of a society, including the assumption of dominant and submissive relationships, may deliberately try to develop distinctly different behavior patterns from mainstream society. Many "new age," egalitarian communities are relatively free from constraints against body contact and place a high value on open, emotional communication between community members (Sommer, 1969). Eye contact is often emphasized as a way of "being in touch" with other people. Stephen Gaskin (1972), the founding leader of a rural community in Tennessee with several hundred people, known as "the Farm," stresses the importance of nonverbal communication, especially eye contact (Gaskin, 1984). Residents of the Farm are characterized by close physical contact and an enjoyment of intimate equal relationships. As one member explained about his life on the Farm: "People always ask what we do for fun here. . . . They don't understand it can be a turn-on to live with other folks, to look into their eyes and let them know you care (Alexander, 1979, p. H4). On Sunday mornings, Gaskin would stand on a stage, surrounded by members of the Farm, and say into the microphone: "If you love someone, look into their eyes and let them know" (Alexander, 1979, p. H4).

Mutual gaze, because of its reciprocal and mutual contact, may be one vehicle for beginning to break down restrictive social roles and the barriers they create between us. As we learn how eye contact affects us, how we use it, and how it is used toward us, we can learn to break down and out of the established roles and in doing so, perhaps recognize the human dignity of the other. The power of the gaze can work this way, too. Even in completely adversary relationships, dignity may be preserved through eye contact—two people realizing that "this is a human being I'm facing." In training for protest demonstrations and direct actions by civil rights workers, antiwar or antinuclear activists, eye contact is invariably emphasized as an instrument of nonviolence; looking into the eyes of the police or government officials reminds the activists that police and officials are "human like you" and gives the police and officials the same opportunity. In this case, eye contact breaks down the roles of police as "pigs" and demonstrators as fanatics, and it gives both sides a means for reducing potential violence. On a larger scale, we may speculate as to how the breakdown of stereotypical roles through social change may increase the possibility of greater equality, and therefore the opportunity for greater intimacy between people. In a relationship of equality, there is no nonverbal expression of dominance and submission, and what we are most likely to find is the free and open interplay of eyes as an expression of caring, friendship, and love.

# NOTES

[1]Thanks to the pioneering work of Charlotte Baker (1976, 1977, in press), the role of eye contact in American Sign Language is being explored. However, more investigation into the functions of nonverbal signals in ASL communication is needed.

[2]In recent years, many physically challenged Americans have been protesting the mainstream view that people with handicaps are incompetent or incapable, disgusting or frightening, and should be hidden from public sight (*Off Our Backs*, 1981). They strenuously object to the basic role they have been assigned—that of being helpless—because the role disregards their individual abilities and needs, and because the role is thoroughly submissive. "Disabled" people are protesting their subordinate role, as are other groups who have also been socially assigned the less powerful position (women, people of color, the elderly, etc.), because they see the role as a barrier to fulfilling or achieving their own needs and goals.

# CHAPTER IV

# Eye Contact and Intimacy

All our lives we juggle the balance between freedom or separateness, on the one hand, and entry or union, on the other. Each of us must have some psychological space within which we are our own masters and into which some may be invited but none must invade. Yet if we insist doggedly on our territorial rights, we run the risk of reducing the exciting contact with the "other" and wasting away. Diminution of contactfulness binds man into loneliness. We see all around us how the reduction of contactfulness can choke man into a condition of personal malaise which festers amid a deadening accumulation of habits, admonitions, and customs. Contact is not just togetherness or joining. It can only happen between separate beings, always requiring independence and always risking capture in the union. At the moment of union, one's fullest sense of his person is swept along into a new creation. I am no longer only me, but me and thee make we. Although me and thee become we in name only, through this naming we gamble with the dissolution of either me or thee. Unless I am experienced in knowing full contact, when I meet you full-eyed, full-bodied, and full-minded, you may become irresistible and engulfing. In contacting you, I wager my independent existence, but only through the contact function can the realization of our identities fully develop. (Pollster & Pollster, 1973, p. 99)

Our society has failed to recognize explicitly the human need for intense interpersonal contact and communication by which individuals experience themselves in relation to other people (Pollster & Pollster, 1973). The ever increasing urbanization, industrialization, and mobility of United States society, along with the breakdown of extended family systems, have caused a depersonalization of our cultural environment and the alienation of people from one another. The ethic of "rugged individualism" has gone too far, undermining close affiliation and mutual support. Our social values tend to devalue intimacy while encouraging competition and alienating types of work. Being "cool" or detached tends to be revered and sought after as a goal in personal development (Hendin, 1975). Unequal power relations between classes, races, and sexes lock us into colder, more distant, and formalized interactions.

As a result of these conditions, many of us have a great, unanswered need for intimacy with other human beings. This need is what attracts many people to cults and sensationalistic groups which promise instant closeness and personal growth. Many people pay exorbitant sums of money for weekend "experiences" which promise to change their lives radically. Our society does not generally offer structures which take care of our needs for closeness. Many of us feel alone, frightened, and unable to contact others in a meaningful way.

In the daily matrix of our social interactions we generally open ourselves to others in routine, shallow ways. Although we are complex beings with many levels of mental, spiritual, and physical reality, only certain socially-prescribed aspects of ourselves are exposed to each other's scrutiny. Generally, we keep close to us what is deep or dark or wounded or passionate—our innermost thoughts and emotions are not casually displayed. However, there are usually some persons in our lives—"intimates" like family, lovers, close friends—with whom we can dare to be whole, to unveil our inner being, and to reveal our truest nature. And the most profound intimacy comes when they reveal their essence to us, as well. Such intimacy can provide the most moving and joyous experiences of our lives.

This chapter's opening quote emphasizes the dialectic between separateness and union in human existence. Persons caught up in the fear of joining with another may avoid a prolonged mutual gaze; they may fear the engulfment of their identity by the eyes of another. Laura Perls (1973) stressed that contact between persons need not be a "fusing" of identities. Rather, she advocated an "elasticity" which allows for the acceptance of oneself and the other as individual, separate people. She described contact as an activity with its own pattern of touching and letting go, "a continuous shuttling or oscillating between me and the other" (p. 7). The eyes can easily become the meeting ground Perls speaks of, a place where two separate identities can meet and communicate. Eye contact can thus facilitate the continuous process of exchange and growth necessary between two individuals.

It is crucial to our development as social creatures that psychologists explore ways to increase interpersonal relatedness. Although the world's literary writers and poets have delved into the experience of the prolonged look in silence, the literature of experimental psychology has unfortunately paid it little attention. Using experiential and phenomenological knowledge, as well as that which has been empirically proven, the present chapter will examine how eye contact functions as one of the most important nonverbal cues in determining the level of intimacy between people; how the relationships between intimacy and eye contact originate in the bond between the infant and her/his caregiver; and generally, how eye contact can be used to further the sharing of feelings and lead to greater trust and closeness in our relationships.

In cooperative and potentially affiliative situations, eye contact intensifies expressions of warmth and empathy (Ellsworth & Carlsmith, 1968). Recognizing this, psychotherapists have begun to emphasize the intimacy value of eye contact. For example, group therapists and sensitivity trainers often ask strangers to engage in eye contact as a way of transcending interpersonal barriers in a group. In the human potential movement, exploration and experimentation have led to nonverbal exercises which increase mutual sensitivity and awareness (Blank, Gottsegen, & Gottsegen, 1971). Touching and eye contact are most valuable for creating intimacy because of the mutual and reciprocal nature of such exchanges.

How is intimacy communicated? In its deepest form, intimacy involves unreserved and mutual verbal and nonverbal communication. Each person freely reveals her/himself through a choice of words or manner of description; through the quality of the voice (perhaps a warm, gentle sound of warmth and acceptance); through touch, near positions, and relaxed, nonthreatening gestures.

The eyes, more than other parts of the body, signal a courageous openness between people through mutual gaze. Generally, the more prolonged the gaze, the more intensely the intimacy is experienced. In addition, the eyes are able to convey many subtle nuances of feeling by their complex capacity for expression. Taken altogether, verbal and nonverbal communication provide the means for intimacy to be expressed and experienced through all our senses. The most intimate moments are perhaps the ones in which all aspects of ourselves are focused on mutual sharing simultaneously, as when making love.

There are many taboos on the forms of expression which create intimacy. Mutual touching has a restricted application outside of familial and erotic relationships, and it is largely confined to handshaking. Conversation is usually possible, but it is a serial exchange rather than a simultaneous one. Nonverbal communication is a more intimate, emotional language than verbal communication. Mutual gazing plays a primary role throughout most interactions and is a reciprocal process particularly suited for enhancing a sense of sharing and thus, intimacy (Heron, 1970).

By definition, intimacy and eye contact are both shared experiences which involve the simultaneous give and take of information between persons. "By the same act in which the observer seeks to know the observed, he surrenders himself to be understood by the other. . . . The eye of a person discloses his own soul when he seeks to uncover that of another" (Simmel, 1921, p. 358). Mutual gazing has a unique capacity to convey people's emotional reactions to one another. Through eyes a human being can "express his excitement at another human being, see that this excitement is contagious and respond to in kind by another, and see that another is also aware of the excitement in both of them and aware of their mutual awareness of their mutual excitement" (Tomkins, 1963, p. 180).

Through eye contact we have a special means for exploring intimacy with others in almost any situation. Although there are indeed taboos on certain kinds of eye behavior, as there are with other forms of intimate expression (Tomkins, 1963), looking is a more subtle behavior and as such may not be so easily noticed by those not directly involved in an interaction. "Stolen glances" can communicate personal interest and generate intimate feelings, much like playing "footsy" under the table (Rubin, 1973).

Many people report that they can experience a strong sense of another person through the eyes. One may look into the eyes of a stranger in a supermarket, during a dramatic audition, across a crowded room—basically at any time—and it can be such a powerful experience that it is remembered for many years afterward. During such revelatory moments of mutual gazing, a type of union occurs in which one feels no barrier between oneself and the other.

In his phenomenological analysis of the gaze, John Heron (1970) emphasized that mutual eye contact involves a process of "reciprocal entry, contact . . . the experience of actual meeting" (p. 245) between two persons. (The only other analogous experience of actual meeting we have is touching.) This reciprocal meeting between two people evokes a sense of intimacy, a feeling of "being" meeting "being." Heron suggested that it is the "luminosity" or "gaze-light" of the eyes which makes us feel that the active presence of another's consciousness is being revealed in us. According to Heron, this gaze-light cannot be reduced to any physical property of the eyes, such as light being reflected from the moist, translucent surface of the cornea. He preferred to describe gaze-light as a "trans-physical" reality that involves mental properties as well as physical ones.

As we gaze into someone's eyes, we experience their luminosity to be "somehow continuous with the very person himself"; the physical properties of the eyes seem to be "interfused with the most intimate activities of consciousness itself" (Heron, 1970, p. 254). We see a greater luminosity in the eyes of musicians, writers, and other artists when they are aroused internally to a heightened state of consciousness and creativity (Heron, 1970). There is also a "streaming" quality to the mutual gaze: "Under optimal conditions of interpersonal encounter, the gaze of the other may be experienced as streaming into my whole being—I am filled out and irradiated by it" (p. 255).

Willingness to engage in a mutual glance has been taken to signify union, closeness, and mutual interaction (Simmel, 1921; Heider, 1958). Searching penetration of the eyes characterizes a human effort to reach another person (Sill, 1900). This is illustrated by Hesse's literary description of an interaction between best friends: "He gazed into my eyes for what seemed an endless time. Slowly he brought his face closer to mine: we almost touched" (1965, p. 171). Ortega y Gasset calls this interpenetration a momentary fleeing from our natural solitude in which tentatively we show "ourselves to the other human being, desiring to give him our life and to receive his" (1957, p. 93). The power of this bond of intimacy is descriptively portrayed by Hesse: "Her great eyes burned close and firmly into mine" (1965, p. 167).

**FIGURE 4-1** A photo from *Eye to Eye: Portraits of Lesbians* by Joan E. Biren (Washington, D.C.: Glad Hag Books, 1979). © JEB (Joan E. Biren) 1979.

## HONESTY

Intimacy involves stripping away the false layers surrounding the self to unveil to others the true essence within. Thus, intimacy involves honesty. The eyes have been famous for their ability to reveal the truth (Reiss, 1969). The act often considered most characteristic of honesty is the act of looking into the eyes of another. In fact, our culture "venerates the act of looking someone straight in the eyes" (Ruesch, 1961, p. 171). Avoidance of gaze is often seen as a manifestation of guilt (Heaton, 1968).

A study by Stass and Willis (1967) gives experimental support for the idea that a look in the eyes is the mark of an honest person. When subjects were told to select partners whom they could trust because the experiment involved money, they chose confederates trained to display a high amount of eye contact. Another study showed that the assessment of a public speaker's sincerity was far greater if s/he looked at her/his audience (Exline & Eldridge, 1967). Exline, Thibaut, Brannon, and Gumpert (1961) have investigated the assumption that honest people always look you in the eye. In their study, students were induced to cheat and then were questioned about their activities during the experiment. At the end of the experiment, those who had been implicated looked less at the interviewer than those who had not cheated.

The direct gaze cannot *always* be taken as a sign of sincerity, however. It was also found in Exline et al.'s study (1961) that those who scored high on a test of Machiavellianism decreased their eye contact significantly less than did the low scorers. Those who want to present a certain kind of image of themselves may consciously exhibit certain kinds of eye behavior. A client in psychotherapy explained why he constantly looked at his therapist saying, "I always look at you because I don't want you to think I'm trying to hide anything by not looking at you."

## LOVE

Of course, intimacy most often occurs in the context of loving relationships. It is in the warm atmosphere of love that we feel most safe to be our true selves. As noted previously, people tend to increase eye contact when they are being positively evaluated, and to reduce eye contact under conditions of negative evaluation. The amount of eye contact persons engage in also has been found to be correlated positively with liking (Exline & Winters, 1965a). Studies have shown that people tend to associate long, reciprocated gazes between same-sex and mixed-sex pairs with longer relationships (Thayer & Schiff, 1974) and, in the case of mixed-sex pairs, long mutual gazes are taken to signify greater sexual interest (Thayer & Schiff, 1977).

In the context of a relationship where there is sexual intimacy, the quality of eye contact may be a way of distinguishing more casual from more intimate sexual relations. Deep communication through mutual gazing may indicate the intensification of an already intimate situation, but research is needed to verify this assumption. As a means of inviting sexual intimacy, the eyes are of major importance (Wright & Joiner, 1974; Symonds, 1972). Their saliency, their unique capability for emotional disclosure, and the common association of eye contact with liking make them well-suited for expression of sexual intent. This is particularly true when there is a prolonged look that violates accepted rules for eye contact and thus conveys a "special" interest.

A client from a foreign country eloquently described a sexual encounter in a bar:

> I met her one of those days when I was mourning the loss of my girlfriend and because my depression was getting out of hand and my suicidal thought was increasing, I decided to get out of home and go to dance. With this decision I also promised myself *never never* never again to be seduced by another's eyes.
>
> At the disco I started to talk, dance, smile, have a good time when suddenly across the salon I saw two very sensual eyes. Those eyes held a very sweet, lonely, passionate, and intriguing look; the kind of look that you cannot classify or fully describe what is behind it. For several minutes I was hooked and aroused by those eyes, I asked her to dance.
>
> She danced so sensual as her eyes looked. I followed her and looked directly into her eyes. I knew that I was heading for trouble. The pace was set. This will be one night to enjoy and seduce. We started to look eyes to eyes setting a sweet challenge to each other. Suddenly it was only her eyes and my eyes were dancing, caressing, kissing with our eyes. What a seductive conversation and my God here we go again seduced by beautiful eyes. When finally we touched, we only follow the road that our eyes have traveled. Later we talked a little because our eyes' conversation was expressed better than our words. So when the moment arrived, we knew how and in which way we were going to spend the night.
>
> We made love and I made love to those beautiful eyes, kissed each centimeter of eyelashes, eyebrows, got immersed in that look and feel like a fly trapped in a spider web, but what a beautiful jail.

Social psychologist Zick Rubin (1970), a pioneer in the investigation of romantic love, found that "strong lovers" (those who scored above the median on his love scale) participated in significantly more mutual gazing than "weak lovers." Although "strong" and "weak" lovers all looked at their lovers a similar amount, what differentiated the two types was the willingness of the "strong" lovers to gaze at each other simultaneously. Thus, the deeper the love is (and, one may presume, the deeper the intimacy), the more eye contact will be evident.

It is important to be aware that Rubin studied romantic love, which may well differ in many ways from filial love, love between siblings, marital love, love between friends, etc. Rubin's conception of romantic love included three components: affiliative and dependent need, a predisposition to help, and an orientation of exclusiveness and absorption. Also, although there are correlations between degree of love as measured by Rubin's scale and amount of eye contact, one cannot assume a causal relation between the two. For example, Rubin explained his results as reflecting the fact that the more "in love" partners were, the more conversation they had and therefore, the more eye contact. Goldstein, Kilroy, and van de Voort (1976) found, however, that there is some evidence for the "lover's gaze," since a significant number of lovers in their study looked at each other more without conversation as compared to strangers in the same experimental condition.

In her study of "limerence" (a word she coined for the experience of being "in-love"), Dorothy Tennov (1980) unreservedly emphasized the importance of mutual gaze. She wrote, "The eyes, as we shall see again and again, are so important in limerence that they, not the genitals or even the heart, may be called the organs of love." In many of the case studies that Tennov cited, looks of all kinds provided fuel for the development of the process of falling in love: brief, tentative, shy looks, secretive glances, flirtatious looks, long, deep mutual gazes, etc.

Beier and Sternberg (1977) observed newlywed couples who self-rated the degree of discord in their marriages. Couples who reported the least disagreement looked at each other the most. (They also touched each other more and sat closer together than couples who reported high disagreements.)

## EMPATHY

When people share a certain intimacy with one another, there are moments when one person is able to "step into the other's shoes" and experience their world from the inside out. We are able to do this through the exercise of the human faculty of empathy, which Lipps (1913) called "feeling oneself into another." This type of intimacy is often referred to with the visual metaphor of replacing one's eyes with those of another person: "I crept inside her skin. She saw the world through my eyes" (Sartre, 1963, p. 34). In the film "Elvira Madigan" (1968), empathy was assumed to characterize love relationships and was referred to with similar eye imagery: "She's made over your eyes. Isn't that what love is? You borrow each other's eyes and you see the world as she sees it." Research has shown that people are most likely to be judged empathic when they engage in mutual gaze (Katz, 1965).

Although it has been assumed in many studies that eye contact leads to intimacy (e.g., Kleinke, 1977) little has been done to test this assumption directly. In an exploratory study by the author (Webbink, 1974), it was hypothesized that three minutes of silent eye contact between a female subject and a female confederate would facilitate intimacy more so than two selected silent control conditions. (The control conditions were each of three minutes duration, but one involved looking at another part of the body, the hand, and the other was an interaction in which no instructions were given other than to maintain silence.) Subjects who made eye contact felt more empathy, positive feeling, and willingness to tell intimate details about their lives to the confederate. If one accepts the postulation that intimacy is composed of the Rogerian (Rogers, 1961) attitudes of empathy, positive regard, and congruence, then this study can be taken to lend support to the idea that eye contact facilitates the creation of intimacy between people. In a study by Ellsworth and Ross (1975), the speakers who had engaged in conversations perceived the situation to be more intimate when the listeners responded with high eye contact.

## GENDER, EYE CONTACT, AND INTIMACY

Although research is not conclusive about the interaction of gender with eye contact and intimacy, the results of some studies suggest that the eye language of intimacy may differ for women and men. For example, Smith and Sheahan (1979) found sex differences in displayed intimacy when they recorded conversations between intimate and nonintimate mixed sex dyads. While females exhibited longer gazes in conversation with an intimate, males gazed longer while interacting with a nonintimate. Ellsworth and Ross (1975) concluded that women respond with intimacy and liking to visual attention; men respond with less intimacy and, for some, less liking. Perhaps, due to sex role socialization, men consider the gaze of another as a challenge to their dominance under certain circumstances (see Chapter II).

## COMPENSATION OR RECIPROCATION

The relationship between eye contact and intimacy as it has been described thus far appears rather simple and direct: It seems that the greater the level of intimacy between persons, the greater will be the level of eye contact. However, as with most generalizations made about eye contact, qualifications to this relationship between eye contact and intimacy must be made.

As mentioned previously, eye contact is part of the larger nonverbal system of communicative immediacy behaviors such as proximity, body orientation, facial expressions, etc. All of these nonverbal behaviors together are used to express intimacy. Therefore, given a certain level of intimacy between two persons, one might find that one nonverbal immediacy behavior, like eye contact, may decrease in its expression of intimacy as another one increases. Thus, an equilibrium at the desired level of intimacy would be maintained (Argyle & Dean, 1965). For example, a strong inverse relationship has been found to exist between frequency and duration of eye contact, on the one hand, and physical proximity, on the other (Argyle & Dean, 1965; Goldberg, Kiesler, & Collins, 1969). This relationship can be explained by the fact that since close distances are physically and psychologically arousing, a person will want to compensate for the added intensity of the powerful mutual gaze (Argyle & Dean, 1965). Also supporting the "equilibrium theory" (Argyle & Dean, 1965) is Kendon's finding (1967) that the more smiling occurs in a conversation the less time will be spent in mutual gaze. Eye contact has also been found to decrease with more intimate topics of conversation (Argyle & Dean, 1965; Amerikaner, 1980), thus indicating a compensatory relationship between eye contact and verbal behavior, as well as with nonverbal. Similarly, Exline et al. (1965) found that subjects looked more at the interviewer during a conversation about their recreational interests than in an embarrassing personal interview.

The level of intimacy with which a person will be comfortable in a social situation can be seen as a function of the balancing of personal and situational approach- and avoidance-forces. The social norms defining particular interactions set the perimeters within which the equilibrium level of intimacy will fall, while the degree of friendliness of the interactants will determine what intimacy level is set within these perimeters (Argyle & Cook, 1976). The personal approach-forces which will encourage eye contact and its concomitant feeling of intimacy include the need for affiliation and the need for positive visual feedback. The tendency to avoid the intimacy of mutual gaze can be created by: fear of being seen, fear of revealing one's inner feelings, and/or fear of becoming aware of negative feedback (Argyle & Dean, 1965).

Equilibrium theory is consonant with J. G. Miller's psychophysiological theory of information overload (1961). According to that approach, it is assumed that there is a limit to the amount of information with which any biosocial system (cell, organism, individual, organization, etc.) can cope. A system faced with more input than it can deal with must do something to reduce the amount, duration, and/or intensity of the stimulation. When an individual's fears of intimacy outweigh her/his need for it, nonverbal immediacy could create a kind of emotional overload requiring the reduction of stimulation through a change in eye contact or physical distance or body orientation, etc. Evidence that avoidance of cognitive overload is linked to avoidance of mutual gaze has been found by Argyle, Ingham, Alkema, and McCallin (1973).

Although equilibrium theory has been recognized as a comprehensive analysis of nonverbal exchange in relation to intimacy (Patterson, 1978b) and has been supported by the results of a number of studies (see Patterson, 1973), there have been several methodological and theoretical shortcomings noted in the psychological literature. Stephenson and Rutter (1970) questioned results which supported the compensatory model. They demonstrated that as the distance between interactants increased, so did the tendency to misperceive looking at the shoulders, ears, or nose as eye contact. Stephenson and Rutter (1970) asserted that Argyle and Dean's findings (1965) about the compensation process were therefore due to the misperception of observers who were judging the occurrence of eye contact from increasing distances. Stephenson and Rutter thus laid a cloud of doubt over Argyle and Dean's (1965) experimental methods and findings that eye contact and proximity are inversely related. Argyle and Ingham (1972), however, have insisted that Stephenson and Rutter's findings are of minimal consequence since, in ordinary interaction, people look at the eyes rather than at other parts of the body.

Patterson (1978b) has developed a comprehensive explanation for discrepancies between studies which have found evidence to support the compensation model and those that have not. He concluded that the kind of awkward interaction between strangers which takes place in the typical research setting facilitated the particular kind of exchange where compensation operates. The typical experimental setting in eye contact research is one in which two strangers, one of whom may be looking steadily at the other's eyes, sit facing each other in an alien and possibly frightening situation (Argyle & Ingham, 1972).

Patterson (1978) emphasized that the "feeling" tone of an interaction is crucial for determining whether a compensatory or a reciprocal process will occur. Basically, when a relationship occurs between strangers, the equilibrium hypothesis applies. Among closer relationships, such as those of good friends, relatives, or lovers, intimacy reciprocation is more likely to occur. For example, when one is fearful or grieving, eye contact or a reassuring touch by a friend or relative may increase intimacy and activate a reciprocal process, causing the *increase* in mutual gaze (Patterson, 1978b). The opposite effect may occur, however, if the fearful or grieving person feels intruded upon by the gaze or touch of a stranger; s/he will thereby avoid mutual gaze or pull away, as the equilibrium theory would predict. Breed (1972) found that increased eye contact was reciprocated by increased eye contact, and in a study by Chapman (1975), children who were sitting closer together while listening to humorous recordings had higher levels of eye contact. In an experiment done in the von Cranach laboratory (1971), the equilibrium hypothesis was also contradicted. When asked to choose a comfortable distance to look at social stimuli, subjects moved nearer to those who looked at them than to those who did not.

Individual differences also mediate patterns of nonverbal intimacy exchange (Patterson, 1978b). Gender differences may interact with results as well, due to the variation in socialization of females and males. For example, the inverse relationship between eye contact and physical proximity was strongest in the mixed sex interactions (Argyle & Dean, 1965). The kind of immediacy behaviors involved will also influence when and how compensation occurs. Two studies by Aiello (1972a, b) demonstrated the complexity of the relationship between proximity and eye contact when the variables of subjects' sex and body orientations are taken into account. Aiello (1972b) varied body orientation and physical distance in a study on sex differences. Confederate gaze was allowed to vary "naturally"; the only constraint was that the confederate had to talk fifty percent of the time. In this study, when male and female subjects were seated at right angles or males were seated face-to-face, subject looking increased linearly with physical distance. However, when females were seated face-to-face at a distance of 6 to 10 feet (as opposed to 2 to 6 feet), this linear relationship did not hold true. Looking began to decrease at 10 feet. The experimenters considered this sex difference in the distance-looking equilibrium formula to be due to differential personal-space boundaries for the sexes.

Aiello (1972a) further clarified these sex-based boundaries in another study in which he found that while for males there was a linear relationship between eye contact and interpersonal distance, for females there was a curvilinear relationship. When conversing at the close and intermediate distances, females returned the interviewer's gaze more, as well as generally looking more, exhibiting longer glances, and looking longer while speaking. At the far distance, where males looked more, the females' looking began to decrease. Contrary to what might be expected by the equilibrium model, females engaged in more mutual gazing with the interviewer at the closest distance and looked more when the interviewer's gazing increased. However, there was an inverse relationship between looking while listening and distance for the females.

Patterson (1976, 1978a, 1981) was able to encompass the diverse aspects of compensatory and reciprocal processes of nonverbal intimacy exchange within an "arousal model of interpersonal intimacy." Many studies have documented that psychological arousal accompanies intimacy behaviors like eye contact and physical proximity (e.g., McBride, King, & James, 1965). Patterson suggested that whether one chooses to respond by compensating for increases in intimacy with a decrease in some component of nonverbal immediacy, or whether one chooses to respond to increased intimacy by a reciprocal increase in one's own immediacy behavior, depends upon how the arousal is experienced. If the arousal is experienced as a negative emotional state, compensatory adjustments will be made. If the arousal is experienced as positive, reciprocal adjustments will ensue. Patterson proposed that

> . . . the many studies showing compensatory adjustments probably involve negative states such as anxiety or some other type of discomfort. That is, when intimacy is increased by a stranger's close approach or stare, that probably increases arousal which, because of the circumstances, becomes negatively labeled. The result is a compensatory adjustment which brings the intimacy back to a more acceptable level. When a similar change in intimacy is initiated by a friend or loved one, in a familiar setting, it may also increase arousal, but now, that arousal becomes positively labeled as liking, love, or happiness. In turn, that experience facilitates a reciprocation of intimacy. (1978a, pp. 5-6)

Whether or not one desires to reduce intimacy or to reciprocate it, changing eye contact is often the most convenient way to alter the intimacy level. Patterson (1973) asserted that "eye contact is undoubtedly much more fluid and flexible in its range of expression than other behaviors" (p. 249).

## THE BOND BETWEEN INFANT AND CAREGIVER

Almost from birth, eye contact is a primary channel for bonding between sighted human beings. Contrary to popular belief, the infant is born with a relatively mature visual system (Rheingold, 1961; Robson, 1967) and reacts with predictable responses to light (Baldwin, 1955; McGinnis, 1930; Shirley, 1933). In a study by Lambers (1981), almost one-third of the observed infants opened their eyes immediately at birth and the majority opened them from within one minute of birth to within twenty minutes of birth. Around two months of age, changes in the strength and direction of pattern preferences occur and visual behavior becomes very similar to what is adopted through life (Rheingold, 1961). In about the third month, the eyes become the first sensory motor system to reach functional maturity in the child, although for boys, perceptual maturity may occur a little later (Moss & Robson, 1968). [The early maturation of female neonates 24 to 60 hours old is evident in the longer duration of their glances, 3.71 seconds, as compared to 2.53 seconds for males (Hittelman & Dickes, 1979).] "Long before an infant can explore his surroundings with hands and feet he is busy exploring it with his eyes" (Fantz, 1961, p. 66).

Visual exploration is the young infant's primary activity as s/he searches for form and movement in the environment (Bruner, 1967; Greenman, 1963; Rheingold, 1961). The visual acts of following and fixation usually appear before the third month and are among the young child's most important achievements (Gesell, Ilg, & Bullis, 1949). Much of the infant's energy is spent looking at the caregiver and following her or his movements around the room with the eyes. Bowlby (1958) suggested that the caregiver's encouragement of these rudimentary responses may be crucial to the infant's psychological development (an idea which will be developed further below). Baker (1967) believed the eyes to be so important to the young infant that he theorized the existence of an ocular stage preceding the oral stage of development. Robson (1967) agreed that the infant's eyes are the most important means of environmental contact.

Findings have shown that infants tend to prefer patterned surfaces and concentric forms which resemble the eyes (Barron, 1968; Bruner, 1967; Fantz, 1958). As early as nine months of age, human infants are more attracted to eyes than to any other visual stimulus (Goren et al., cited in Freedman, 1979). Magaziner (1974, cited in Freedman, 1979) found that crying newborns ceased crying when they were picked up and turned toward the caregiver. The newborns initiated eye contact by opening their eyes to look at the caregiver's face and fixating on the caregiver's eyes. Children from 4 to 7½ months old in a French daycare center were placed in infant seats facing each other at a ninety degree angle. Of the forty-four responses to the other child that were coded, the only consistent behavior was the initial fixation of the other child's face (Rouchouse, 1978).

In the feeding situation, as the baby gradually begins sucking with open eyes, eye contact often becomes as important to the child as the food itself (Bruner, 1968). Through eye contact, the infant relates not only to the food, but to the person who gives the food (Klein, Heimann, Isaacs, & Rivieres, 1952). Robson (1967) found many mothers reported that eye contact dominated the feeding situation. On the basis of films of nursing babies, Spitz and Wolf (1946) noted: "The nursing infant does not remove for an instant its eyes from its mother's face until it falls asleep at the breast, satiated" (p. 109). The baby may even stop feeding if the mother looks away (Heaton, 1968).

To test eye importance to the young child, Shapiro and Stine (1965) collected figure drawings of 3- and 4-year-old children. In the sample of children under 4 years of age, 89% drew eyes, while 22% drew the mouth. These psychologists also found that the eyes were represented independently of both nose and mouth, which tended to be fused together and drawn at a later age. They concluded that "this stimulus configuration is selectively attended to as the 'locus vitae' of the infant's primary caretakers, and as an important organizer of his perceptual world" (Robson, 1967, p. 17).

Buhler and Hetzer (1928) stated in their early work that 3- and 4-month-old infants responded to the eyes with a smile, and that the facial gestalt as a whole did not gain importance until the age of 5 months. Kaila (1932) found the releasing stimulus to be the whole eye region. Spitz and Wolf (1946) extended this region to include part of the forehead and nose. Ahrens (1954) found that the smiling response was elicited by as many as six eye-sized dots in children up to 2 months old. In the second month, two horizontal eye-sized dots became the best releasers. At the age of 4 months, when given a mirror, the child chiefly regards the reflection of her/his own eyes (Gesell, Ilg, & Ames, 1956). As the infant develops, the whole human face becomes an optimal releaser. From these studies, it can be concluded that this child responds not to the human face or facial expression, but to a facial configuration in which the eyes are the most important elements.

The tendency of human infants to orient their eyes toward the eyes of another has great import for the child's survival. The searching eyes of an infant both elicit and reward the caregiver's attention and affection. This effect is crucial for a being who enters the world in such a helpless state (Magaziner, 1974, cited in Freedman, 1979).

Eye contact has been found to be even more important than the smile when it comes to the mother's response to her infant. Perhaps this is due to the fact that without eye-to-eye contact accompanying a smile, the mother cannot be certain that the smile is meant for her (Hutt & Ounsted, 1966). Robson (1967) and Wolff (1963) reported that mothers felt estranged from their infants until about 4 weeks of age, at which time the babies acquired the ability to maintain true eye-to-eye contact (up until this age, the child usually appears to be focusing at a point somewhere behind an observer's head). Then the mothers felt recognized, more intimate with and more affectionate toward their babies, although they often were not aware of the cause for the change in feeling. Three of the mothers in Wolff's study began to spend much more time with their infants during the fourth week. Robson (1967) proposed that eye contact should be added to Bowlby's (1958) list of innate releasers of maternal caretaking behaviors. (Other releasers on the list include crying, smiling, visual and physical following, clinging, and sucking.) Because the baby has no inhibitions about intimacy, eye contact with one's infant can be one of the most intense and deep forms of communication that exists. One can literally experience the infant's eyes pouring into one's own.

Of course, other dynamics develop in the case of blind infants and their mothers who are deprived of a visual bond. Fraiberg (1974) has conducted longitudinal studies of the interactions between mothers and their infants who are blind from birth. As researchers and clinicians who have worked with blind children for several years, Fraiberg and associates "have never overcome this sense of something vital missing in the social exchange" (1974, p. 217). Fraiberg (1974) advises mothers to use much vocalization and physical contact to cultivate their relationship with their blind children. Blind infants who lack such exchanges often show damaged object and human relationships later in life.

Mutual gaze is crucial to both mother and infant in the development of an emotional bond, a sense of connection, and closeness between them. As noted above, for the mother, the gaze of the child into her eyes serves as an innate releaser of maternal caregiving and affection. For the child, the gaze of the mother into her/his own eyes is the first truly reciprocal experience of union with another being. For this reason it is probable that eye contact between the infant and caregiver is a precursor for the future relationships and emotional development of the child (Klein et al., 1952; Robson, 1967). Analysis of the important role of eye contact in the caregiver/infant relationship will surely help us to better understand why eye contact has such a major role in the communication of love, empathy, and other components of intimacy in human interaction.

FIGURE 4-2 Author and her baby making eye contact.

Although little attention generally has been paid to developmental issues in the field of nonverbal behavior, there has been some investigation of eye contact in this regard. However, most studies have focused only on the child. It has been a fairly recent phenomenon to study how the caregiver's behavior is dependent on the infant's signals (Vine, 1973). Lewis and Rosenbaum asserted that: "Not only is the infant or child influenced by its social, political, economic, and biological world, but in fact the child itself influences its world in turn" (1974, p. xv). Robin (1980) found that, in the first week of life, the infant staring into the mother's face elicited an increase in her communication activities with her child and that these activities often ceased as soon as the child closed her/his eyes. Stern (1974) applied an *interactive* model to the mutual gazing patterns between caregivers and children, viewing the exchanges between them as ones in which "influences flow in both directions" (1974, p. 187). For the infant, the eyes are the first channel available for functioning at a level comparable to the caregiver's. Thus, mutual gazing provides the first experience of perceptual equality and mutuality with another human being. Stern's (1974) interactive model supports Fehr and Exline's (in press) proposition that the subtle sex differences in average gaze level (due to the earlier maturation of female infants) may have an impact on the caregiver's handling of the infant, thus promoting the more visually affiliative behavior found in older female children and adult females.

Through the eyes the child has a means of influencing another, much as s/he is influenced. Control of perceptual input through the eyes allows the infant to maintain her/his internal state of stimulation within a desired range (Stechler & Carpenter, 1967; Stechler & Latz, 1966; Walters & Parke, 1965). By either meeting another's gaze or by gaze aversion, the infant is able to control the amount of visual social contact to which s/he is exposed. Through gaze aversion, a child is often able to terminate an intrusive behavior on the part of the caregiver. Farren, Herschbiel, and Jay (1980) found that for mother–child pairs with children ranging from 6 to 36 months, mutual gaze was most often initiated by the mother and terminated by the child. Leavitt and Donovan (1979) found that the averted gaze of an infant was arousing to mothers. In a study by Brown (1972), the low-gaze babies received significantly more maternal care (and were significantly more irritable) than the high-gaze group. The mother's sensitivity to gaze aversion allows her infant to take part in the regulation of social interaction and thus to have some measure of control in her/his environment.

Often the mother's gaze elicits a smiling response in the child, with the child's smiling response subsequently causing the mother to smile back (Gerwirtz & Gerwirtz, 1969). Mutual gazing thus elicits social reinforcement (the mother's smile) for the child's behavior sequences and for mutual gazing itself. This points again to the primary role that eye communication plays during the preverbal years of human life, the time when nonverbal communication must transmit all meaning from the infant to caregiver and vice versa (Vine, 1973).

Although there is no research on the subjective experiences of children during eye contact, one might hypothesize that some of the most important social experiences of the child's preverbal life takes place in mutual gazing. If one spends any time with children or infants their unique capability for prolonged gazes that may be described as "deep" or "meaningful" becomes apparent. Although a child may not be able to explain in words how it feels to spend long periods of time gazing into her/his mother's eyes, it is probable that such moments are infused with emotions of intimate sharing. The experience of such moments can, in turn, create a foundation of trust which makes further intimacy possible.

Through eye contact a child may be able to experience empathically the emotions of a parent. For example, because of the facial expression accompanying eye contact, the child may recognize pain in the parent's eyes. Feeling the parent's pain through their eyes at an early age may be very difficult to talk about, even after language is acquired, because it is very difficult to put nonverbal experience into words. Although there has been little research into the capacity of infants to differentiate emotions in the caretaker's face, both Benjamin (1963) and Meili (1957) noted that after the fourth month, babies are very sensitive to changes in the upper portion of the face. A grave look on a mother's face can produce a fear reaction in her infant. Seeing pain in a parent's eyes could cause a child to avoid mutual gazing.

Marked variations in eye contact patterns among infants were observed by Robson (1967). Some infants were visually alert, making attempts to engage the mother in a mutual gaze and appearing quite engrossed when contact was achieved. Others made contact but without the apparent fascination of the first group. Still others seemed to avoid their mother's gaze.

What determines eye contact patterns in infants? A study by Moss and Robson (1968) found a relationship between the attitudes of expectant mothers and their later patterns of eye contact with their 1-month-old infants. The more positive maternal attitudes were during pregnancy, the higher were the amounts of mutual facial regard observed later. Kulman (1962, cited in Katz, 1965) found that during feeding, the percentage of time the baby looked at her/his mother was matched by the percentage of time the mother looked at the baby. On the other hand, Hutt and Ounsted (1966) suggested that there may be cases where, for some reason, the infant does not provide the releasing stimulus of eye contact which strengthens the mother–infant bond. From the perspective of the interactive model, it seems likely that neither caregiver nor child can be considered solely responsible for the gaze patterns between them.

After extensive research, Robson (1967) concluded that "the nature of the eye contact between a mother and her baby . . . conveys the intimacy or 'distance' characteristic of their relationship as a whole" (p. 18), a conclusion which has been supported by more recent studies (e.g., Bampton, Jones, & Mancini, 1981). Robson then theorized that, because mutual gazing between mother and child is the foundation of early communication and human bonding, early eye contact patterns can determine not only how the child will utilize eye contact as an adult, but may also affect the nature of her/his relationships in general (Jaffe, Stern, & Perry, 1973; Robson, Pederson, & Moss, 1969).

For example, the visual interaction between mother and child in the first three months of life is related to the child's later readiness to interact with strangers; i.e., the more stimulation the child has received, the more prepared s/he is to interact comfortably with others outside the family (Moss, Robson, & Pederson, 1969). Likewise, "attachment disturbances" in human relationships probably will be accompanied by "enduring deviations in eye-to-eye contact," according to Robson (1967, p. 22).

Robson (1967) emphasized that mutual gazing between mothers and their infants is crucial to the creation of a bond of attachment between them. He theorized that perfectly adequate caretaking behaviors like bathing, changing, and dressing are not enough to develop a "face-tie" between mother and child unless they are accompanied by the initiation of mutual gazing on the part of the mother. Robson also pointed out that it is not merely the presence of eye contact that is important, but also the quality of such eye interactions.

Ruesch (1972), in a discussion of nonverbal behavior in general, also emphasized the importance of the caregiver's paying attention to the nonverbal signals of a child. Parents who respond only to a child's verbal signals may contribute to "nonverbal retardation"; i.e., "his appreciation of space, beauty, and movement lags behind the skill displayed with verbal or digital symbols. Such a child may become a verbal genius while he remains nonverbally feebleminded" (Ruesch, 1972, p. 362). It is important that the child obtain a complete response to all of his/her expressions.[1] The habitual, largely unconscious nonverbal acts of adulthood are patterned through the social interactions which occur in childhood (von Cranach & Vine, 1973).

Observations of white, middle-class mothers and their 3- to 4-month-old babies at play revealed that maternal facial and vocal behaviors were elicited not by the presence of her child in general, but specifically by the child's gaze at her (Stern, 1974). When compared to adult interactions, the gaze behavior between a mother and her child might be labeled "deviant." Eye contact between mothers and their children often lasted over 30 seconds in Stern's study; mutual glances of such length rarely occur "between adults in this culture, except between lovers and people about to fight" (Scheflen, cited in Stern, 1974). Whereas the mother actually talks for both herself and her infant while the infant listens and rarely vocalizes, the mother strays from our adult cultural pattern by taking on the role of an almost constantly gazing "listener," acting as if the infant were talking (Stern, 1974).

The occurrence of a very special kind of eye contact has been observed between mothers and their children: moments of extended, loving gazes which take place in silence. As Stern (1974) described such gazes, they often have "the aura of a very quiet magic moment" (p. 209). Such moments may be experienced as a sense of a deep, almost "psychic" intimacy with a child (White, 1984).

Adults seem to reexperience such moments from their childhood years in the looks which occur between lovers. Tomkins (1963) explained, "Indeed, many of us fall in love with those into whose eyes we have permitted ourselves to look and by whose eyes we have let ourselves be seen. This love is romantic because it is continuous with the period before the individual lovers know shame. They not only return to baby talk, but even more important they return to baby looking" (p. 172).

There is a small but growing body of research on the relationship of age to duration and frequency of eye contact. In a preliminary study by the author (1974), trends in eye contact behavior varied according to age (and sex). Older children looked more and had more instances of eye contact than younger children. Ashear and Snortum (1971) found that eye contact with a steadily-gazing adult female peaked at kindergarten and second grade and reached a low point at fifth grade. Levine and Sutton-Smith (1973) found that gazing with a same-sex peer increased between the ages of 4 and 9, decreased between 10 and 12, and then increased to reach the highest point for adults. Although there was some post hoc evidence that the drop in gazing between 10 and 12 years of age may have been due to self-consciousness, Levine and Sutton-Smith (1973) were unable to account for all the age changes in gazing by any single factor.

A comprehensive review of the literature on social gaze by Fehr and Exline (in press) found that several studies have documented the tendency of female children to gaze at others more than male children (Ashear & Snortum, 1971; Kleinke, Desautels, & Knapp, 1977; Levine & Sutton-Smith, 1973; Russo, 1975; Vlietstra & Manske, 1981). Levine and Sutton-Smith (1973) observed the proportion of total gaze, mutual gaze, looking while listening, and looking while speaking. They found that girls exhibited higher gaze levels in the measures of mutual gaze and looking while speaking.

Only a few researchers have studied children's responses to and interpretations of gaze. Again, sex differences have been observed. Kleinke et al. (1977) found that girls (3 to 5 years of age) preferred an adult female when she gazed steadily, while boys liked the confederate less. In a study by Vliestra and Manske (1981), 4- and 5-year-olds interpreted pictures of children interacting with either steadily gazing adults or with adults who looked away. Both boys and girls interpreted the gazing adult as more approving. However, when the children who had viewed the pictures of the gazing adult actually interacted with a steadily gazing adult, the boys gave a more disapproving interpretation than the girls and girls gave more approving interpretations than boys. Also, the more boys had glanced at the adults, the lower were their approval ratings.

Vaughn and Waters (1981) broke new ground in the study of visual behavior in children when they examined the relationship of visual attention to social organization among preschoolers. Daily observations of the children during a three-term school year revealed the existence of a stable attention structure or hierarchy. High-ranking children received more visual attention from others than would be expected by chance. Because visual attention ranks were positively correlated with measures of conversation with peers, interactive play, and positive sociometric choice, the authors concluded that the structure of social organization in these preschool children was based on social competence. Among preschool children, social competence attracted visual attention.

Although the studies above provide us with a new wealth of information about the patterns of children's eye contact, certain considerations must be kept in mind when examining these findings. Anita Woolfolk (1981) pointed out that studies of adult–child communication usually use adult observers to judge the nonverbal behavior of children, and do so assuming that the child is perceiving and responding to the same cues as the adult has identified. Woolfolk cautioned against this easy assumption, since no one has yet demonstrated that adults are the most accurate judges of children's nonverbal communication or that adult perceptions necessarily correspond to the child subject.

## AREAS FOR FUTURE RESEARCH

Since many psychologists have assumed that early eye contact experiences predispose later eye contact and intimacy patterns, it seems quite important that investigators expand their exploration of the development of eye contact during the course of a child's life. Information from such investigations would enhance our understanding of eye contact, intimacy, and other variables. Until such research has been accomplished, we should not accept theories about the relationship between early eye contact experiences and adult behavior without some reservation. In addition, there is much more room for investigating the relationship between eye contact and intimacy. For example, one could attempt to distinguish the levels of intimacy which take place in eye contact. Just as there are many ways in which one person can experience and respond to the touch of another (e.g., receiving a kiss coldly or returning it passionately), so eye contact can range from the response of a blank stare to a deep gaze which penetrates the other's "soul."

Qualitative differences in eye contact, perhaps the most fascinating and fruitful area of exploration, have been ignored to a large extent in the psychological literature. Eye contact has been generally defined in a gross fashion and inferences are made on this behavior without distinguishing the type of contact which occurred. Perhaps investigators could discover the important questions to ask about intimacy and eye contact by first looking carefully at the intimate interactions between two people, and then studying larger samples. For example, it would be most interesting to study two people who are passionately in love with each other, measuring their GSR and EEG, watching their eye responses to each other as it was coordinated with their conversations. Hypothetically, one might find that when they say, "I love you," their GSR will change and something would happen to their eyes. Perhaps their eyes would open more or appear more teary. Later one could check whether these results are replicated in larger populations.

If we could discover what it is that "happens" in eye contact, we could better help people who are having problems with intimacy. If we could refine our knowledge about eye contact and intimacy, perhaps therapists could use it even more effectively in the process of creating a therapeutic relationship. After all, the therapy experience is meant to provide an opportunity for the client to reveal her/his self so that s/he can explore the inner world beneath defenses. If therapists had this detailed information, they could use eye contact more effectively to encourage vulnerability and self-awareness—they could use it gently to reach someone deeply.

## CONCLUSION

In our quest as social creatures for moments of union with others, eye contact plays a major role. Through its simultaneous and reciprocal nature, eye contact allows a mutual openness and sharing between persons that is comparable only to touching. Research has shown that honesty, love, and empathy are often identified with mutual gaze.

Eye contact has been found to vary according to the level of intimacy between persons. In situations where nonverbal immediacy is experienced as negative arousal, people will probably compensate for increases in nonverbal intimacy by decreasing eye contact. Where increasing intimacy provides positive arousal, immediacy may be reciprocated by an increase in eye contact.

Psychologists have emphasized the importance of eye contact between children and their primary caregivers. Some have asserted that adult social behavior is related to early eye contact experiences between mother and child. At the very least, eye contact plays a major role in the development of an emotional bond between infant and caregiver. The patterns of mother–infant interactions have been shown to vary according to the behavior of both the mothers and their children.

The relationship between age and eye contact patterns needs greater clarification in the experimental literature. Furthermore, the quality of eye contact has lacked any serious attention. Despite gaps in the experimental literature, however, some research and much phenomenological evidence points to a strong relationship between eye contact and intimacy. For those who would like to engender greater intimacy in their social interactions, awareness of the role eye contact plays is invaluable.

## NOTE

[1]The caregiver's responses should follow the infant's signals within a period of time short enough to allow the infant to associate her/his actions with her/his caregiver's responses. The caregiver's facial expressions also should be modulated to fit the baby's needs at the time, "so as to reassure rather than disrupt, or quiet rather than prolong distress" (Robson, 1967, p. 21). Contingent behavior on the part of the mother may be, according to Robson, a basis for establishing both trust and the subjective sense of meaning in the child. Stern, Beebe, Jaffe, and Benett (1977) suggested that the temporal patterning of caregiver/infant behaviors is more important than the modality utilized. "The caregiver . . . in trying to care for and engage the infant, creates themes and variations of sound and movement which the infant's mental processes will gradually retranspose into classes of human acts of caring and engaging" (p. 193).

# CHAPTER V

# Clinical Aspects
# of Eye Contact

Common experience tells us that a person's eyes can reveal much about her or his immediate state of being. The "happy" individual has clear, bright eyes that look openly into the eyes of another. When someone is unhappy or ill, her/his eyes may appear lifeless and s/he may have difficulty maintaining mutual gaze. In cultures where emotional problems were considered to be caused by demon possession, the expression of the eyes was thought to be the most revealing diagnostic sign of "health" or "insanity" (Greenacre, 1926). In 1908, Diefendorf and Dodge stated that they could detect "insanity" in people by studying their eyes in photographs. Although modern science does not make that claim, the individual's gaze is an important index of her/his psychological "strength" (Riemer, 1955).

In recent years, diagnosticians and therapists have become increasingly aware of the critical value of eye contact in the clinical setting, above and beyond its ordinary function in human relationships (Ratnavale, 1984). They have discovered that eye contact can be a means for not only establishing a relationship but for gaining an empathic understanding of the individual as well. Eye contact can be utilized as a means for stimulating or intensifying emotion, as a specific focus of the client's[1] behavior (e.g., examining and changing gaze patterns to encourage healthy interpersonal relations), and as an integral part of psychotherapy (Webbink, 1980 & unpublished).

Within a specific culture, variations of eye contact and looking behavior can provide important clues about a person. The direction of eye movement from left to right or right to left was found in the late 1960s by Merle Day (1967) to be indicative of different emotional states. Furthermore, the direction of eye movements both vertically and horizontally were more recently found in a method called neuro-linguistic programming to correlate with the person's primary "representational system," that is, visual, auditory, or kinesthetic. To give an illustration, if a person primarily uses a visual representational system, and uses phrases such as, "the way I see it"; "I view it as"; "it looks to me like"; "do you get the picture?," s/he might tend to look up and to the left. The person is "accessing" things by visual memory (Bandler & Grinder, 1979).[2]

Individuals develop distinctive nonverbal styles which have been found to be highly stable over time and across situations (Kearns, 1976). Every person's life experiences contain elements that contribute to individual differences in nonverbal style, much as unique experiences help create distinctive personalities. Nonverbal styles correlate with personality attributes. For example, British researchers Lamb and Turner (1969) and Ramsden (1976) have observed that movement styles can indicate a person's flexibility, decision-making abilities, communication skills, potential for taking initiative, capacity for assessing problems, etc. Certain individual movement characteristics also have been found to be significantly correlated with forms of "psychopathology" (Davis, 1970; Davis, 1979b).

Individuals vary greatly from one another in the quantity and manner of their looking behavior. Kendon (1967), in England, found that the amount of time the people in his sample spent looking ranged from 28 to 70%. (For stressful interviews, Nielsen's 1962 sample, in Denmark, showed variations from 8 to 73%.) Despite the influences of "gaze consensus" (see Chapter I) and other situational and cultural factors, each person engages in similar amounts of eye contact during interactions with different people. Research findings are thus consonant with widely held beliefs which associate eyes with individual identity and which consider the eyes to mirror the "soul." Every person exhibits her/his own unique pattern of eye contact, and some researchers have found evidence to support the idea that "the ocular response . . . betrays or indicates underlying personality structure" (Libby & Yaklevich, 1973, p. 203).

Eye contact patterns are particularly indicative of people's ability to relate on an interpersonal level. For example, Cegala, Alexander, and Sokuvitz (1979) found high levels of gaze to be related to higher levels of self-reported involvement in an interaction. The degree to which a person is open to social interaction seems to differ among individuals according to whether a person's primary motivations are related to affiliative needs or to needs for control in the environment and/or in relationships (Libby & Yaklevich, 1973).

Exline (1962) reported some informal observations of people who had been divided into affiliative versus achievement-oriented groups.

Although his observations in this study were verified only in an ad hoc fashion, he offered a very descriptive picture of the differences in the gaze behavior of affiliators and achievers:

> Whenever a member of an affiliative group spoke, whether male or female, the speaker would evenly distribute a considerable amount of visual attention around the group. When making a general statement, he or she would first rest his gaze upon one person, then another, then the third, and fourth, until the whole group had been encompassed in a rather stately progression. Equally striking were the behaviors of the affiliative listeners. They appeared to rivet their visual attention upon the face of the speaker, giving the observer the impression that he was the relatively constant focus of four pairs of orbs.
>
> If, in the general discussion, one affiliator spoke to a point made by another, the two would first engage each other in a mutual glance, usually broken by the speaker to look at the others who sat watching him.
>
> On the other hand, the relatively more achievement-oriented speaker was more prone to give an initial quick sweep of the group, then focus his attention on notes in hand, or on the middle distance. Occasionally he would sweep the group again before returning to his notes or briefly engage one person who may have inadvertently caught his eye. Neither did the achievement-oriented listeners look steadily at the speaker. They too would peruse their notes, look into the distance, or occasionally look at the speaker when his visual attention was elsewhere. (p. 164)

Mutual gaze helps people become emotionally closer to one another because of its ability to make people feel their existence is recognized; because it allows sharing of emotional and other kinds of information; because it is often reinforcing, etc. Thus, it follows that people who are in greater need of affiliation will participate in more mutual gazing (Argyle, 1975).[3]

Several studies have shown that individuals who are considered to be more dependent (i.e., rely on others for help and approval) exhibit higher levels of eye contact (Exline, 1963; Mobbs, 1968; Nevill, 1974). A person who is motivated for dependency has a strong need to attend to the environment. One study found that dependent subjects looked more under conditions of low reinforcement than of high reinforcement (Exline & Messick, 1967). The dependent person may look more when s/he is not being reinforced because s/he has a high desire to please the other and is thus attentive visually in order to discover the effect of her/his performance.

Similarly, Efran and Broughton (1966) found that those who are approval-dependent in mutual relationships have greater eye contact than those who are not. Later, this finding was lent further support by Pellegrini and associates (1970) who induced the desire for approval in subjects, and found that those seeking approval made more eye contact than a control group and those avoiding approval. Argyle (1973) noted that those who are more concerned about the reactions of others will make shorter utterances and look up more. Some data also show that people categorized as "field dependent" take part in more looking. "Field dependent" people are more affected by their surroundings and, therefore, probably need "to keep a close eye on people in it" (Argyle, 1973, p. 108).

Individuals who have been described as more affection- and inclusion-oriented also have higher levels of eye contact (Exline, Gray, & Schuette, 1965). In addition, high-nurturant people look more than those who are low-nurturant (Libby & Yaklevich, 1973). If we consider nurturant people to be characterized by a desire to give of themselves to others, then this "evidently includes giving with the eyes" (Libby & Yaklevich, 1973, p. 203). Marks (1971) found that persons who "more readily engage in eye contact" were more empathetic (a characteristic associated with nurturance) than eye contact avoiders.

Persons who have a great need for control in their lives, those who have power over others, and those who have obtained high-status positions exhibit characteristic patterns of eye contact. Briefly, it has been found that, except in situations where there is a challenge to their authority, such persons look less than others who have less power or who are less control-oriented. (Eye contact patterns for dominant vs. submissive individuals are described in detail in Chapter II.) Low control-oriented individuals are likely to behave similarly to those who are highly motivated for dependency, exhibiting higher levels of eye contact.

Persons who are motivated to control others and who are not in need of social rewards have been defined in the psychological literature as "Machiavellians" (Christie & Geis, 1968). Such persons are effective in coldly manipulating others through psychological, rather than physical, means. Although looking behavior often operates at a lower level of consciousness than does ordinary speech, it has been observed that some people, whether or not they are aware of it, have a high degree of control over their eye muscles (Reimer, 1955) and that Machiavellians have cultivated the ability to control eye behavior in order to manipulate others. While most people will look less after they have been involved in an unethical act than they did before the act, subjects who had scored high on the Christie-Geis Machiavellianism Scale did not decrease the amount of their gaze (Exline, Thibaut, Hickey, & Gumpert, 1970).

Observation of eye contact can play a significant role in diagnosis and is especially suited for that purpose in cases exhibiting disturbed, minimal, or nonexistent interpersonal relationships (Harris, 1978).

During the first critical interview with a client, the clinician may note the amount of eye contact as well as its qualitative aspects. A diagnostician "must be alert to every variation in level, intensity, and nuance of eye contact while remaining very mindful of the fact that too great a focus on eye contact could significantly influence such behavior" (Ratnavale, 1984, p. 9). Ellsworth and Ludwig (1972) asserted that people with severe psychological problems deviate, sometimes strikingly, from normal gaze patterns. For example, clients in panic states may be hypervigilant, restlessly scanning their environment. Persons with toxic or manic conditions may show extreme distractibility (Ratnavale, 1984). Often deviations are manifested by a refusal to maintain or even establish eye contact at all (Rutter & Stephenson, 1972).

The individual's reaction to the gaze of another person can be another indication of psychological problems. Laing (1960) pointed out that an extreme reaction to a look is characteristic of the person who either dreads or cherishes being looked at. Persons diagnosed as paranoid schizophrenic, for example, may perceive a look as accusatory or penetrating (Laing, 1960). Thus, one way for this individual to maintain psychological equilibrium is through an active avoidance of eye contact. Likewise, persons with low self-esteem exhibit patterns of reduced eye contact, perhaps due to a need to avoid negative cues which might confirm their "failures" or the validity of their self-concept (Fugita, Agle, Newman, and Walfish, 1977).

Gaze aversion is also seen in other individuals who have a drastically reduced capacity to relate interpersonally. Since diminished (or nonexistent) interpersonal relationships are relatively characteristic of childhood autism, schizophrenia, and depression, decreased eye contact may be particularly suited as a diagnostic tool in such cases (Argyle & Kendon, 1967; Rutter, 1973).

## CHILDHOOD AUTISM

Ellsworth and Ludwig's (1972) observation of deviant patterns of eye contact is exemplified most dramatically in the behavior of autistic children. Researchers have reported that a refusal to maintain eye-to-eye contact is a major feature of infantile psychosis (Rimland, 1963; Hutt & Ounsted, 1966; Matsusaka, 1972). Gaze avoidance is particularly characteristic of autistic children during the toddler stage of development (Kolvin, 1972). Bernard Rimland (1963) observed that the autistic child adamantly avoids eye contact, even to the point of struggling to move the head when held face-to-face with an adult. Hutt and Ounsted (1966) described the typical behavior of an autist held in the arms of an adult: "The child keeps its face averted from the adult's face, and any attempt to make it fixate upon the adult provokes the child to shield its eyes with its hands" (p. 347). These children are often said to "look through" others. Their expression resembles that of people who have been subjected to prolonged sensory deprivation.

Furthermore, in an experiment involving five cardboard faces, one of which was a happy human face and another sad, Hutt and Ounsted (1966) found that autistic children visually avoided the happy face. The experimental design was repeated in order to determine whether it was the face with only a smiling mouth or the face with only eyes that evoked the negative reaction. They found that the children avoided the face with eyes more frequently than the face with a mouth. The research of O'Connor and Hermelin (1967), however, appears to contradict these findings. In their study comparing autistic, mentally retarded, and normal children, all three groups took more time looking at a photograph of a face with eyes than at any other display. In another presentation, the psychotic and normal children spent the same amount of time gazing at an open-eyed face as at the same face with its eyes closed.

It has been suggested that autistic children are in a state of high behavioral and physiological arousal (Rimland, 1963; Hutt, Hutt, & Ounsted, 1965; Hutt & Ounsted, 1966). This state may originate in various biochemical disorders; however, the specific nature of these has yet to be determined. According to Rimland (1963), these children have a defect in the reticular activating system which results in chaotic confusion for the child. Whatever the cause, autistic children appear to require a reduction of massive perceptual stimuli, especially highly arousing stimuli, such as eye contact (Rimland, 1963). If this is the case, then gaze aversion would effectively reduce arousal. It may even be that gaze aversion is a biological component of such arousal (Hutt & Ounsted, 1966). In a study by Jurich and Jurich (1974), "percent of no eye contact" was highly correlated with people's sweatprints and other measurements of anxiety (Jurich & Jurich, 1974).[4] Increases in eyeblink frequency often accompany increases in psychophysiological activity, especially when elevated arousal is due to unpleasant experiences (Tecce & Cole, 1976). Gaze aversion also occurs at high levels of intimacy and with close proximity (Argyle & Cook, 1976), which further indicates the role of gaze aversion in arousal reduction.

The avoidance of eye contact may have additional benefit for the autist. By observing autistic children in a playroom with other children, Hutt and Vaizey (1966) found not only that the autists displayed little or no aggression but also that they were never attacked by the other children. It may be that the autists' submissive behavior, exemplified by their gaze avoidance, acts to appease the other children and effectively inhibit aggressive behavior (Hutt & Vaizey, 1966).


## ADULT SCHIZOPHRENIA

According to Reimer (1955), adult schizophrenics also manifest abnormal eye behavior, most notably gaze aversion, and this aversion is characteristic of an individual's avoidance of interpersonal relationships. It appears that the amount of eye contact exhibited by the individual is related to the severity of the disturbance. Lefcourt, Rosenberg, Buckspan, and Steffy (1967) found that in a 4-minute period, acute schizophrenics engaged in more eye contact than did chronic schizophrenics. Reimer (1949) reported on some schizophrenic patients who found it extremely difficult to maintain eye contact. He noted that averted gaze generally indicates a state of fear and shame. Laing (1960) stated that schizophrenics are overwhelmed by rejection and do not wish to be seen. The withdrawal typical of these individuals is an extreme form of shyness and modesty. Clancy (1963) offered the hypothesis that this avoidance symbolically represents an inability to face situations or a fear of looking at others. Behavioral guardedness may also be indicated by extensive visual scanning, as in the schizophrenic whose eyes are always moving from object to object (Schooler & Silverman, 1969). Tecce, Savignano-Bowman, and Cole (1978) have found that psychological disturbance is reflected in increases in eyeblink frequency. Both borderline and chronic schizophrenics have been found to have higher eyeblink rates than young (as opposed to elderly) "normals" (Tecce, 1978).

According to Wilhelm Reich (1972), when a schizophrenic does gaze at another person, the expression of the eyes may be blank. This look has been captured by Paul Christensen's (1978) description: ". . . her eyes/say nothing as if the lights/were out beyond the first/sharp turn of her mind." Reich claimed that the "blank" look is caused by contraction of certain muscles around the eyes. He described the schizophrenic person's eyes as "peeping out . . . from a rigid mask. The patient is not capable of opening the eyes wide as if to imitate fear" (p. 369). Reichian therapists note that, in some cases, this "eye block" is central to schizophrenia (Dew, 1970; Konia, 1970). Many of these persons also lose their capacity to shed tears (Reich, 1972) and tend not to have headaches because the pain does not penetrate the ocular armor (Dew, 1970).

Schizophrenics often fail to provide verbal information about their relationships and many psychiatric care workers feel that the information parents provide is unreliable. Therefore, Harris (1978) decided to observe clients' mutual gazing patterns for clues about their relationships. Harris found that the schizophrenics (who were all males in this study) made themselves significantly more available for eye contact with their mothers rather than with their fathers, authority figures, or peers. Thus it appeared that male schizophrenics were more willing to interact with their mothers than anyone else. For Harris, these patterns of eye contact supported the idea that, "Mother still defines for the schizophrenic what is acceptable, rational, and important" (p. 90).

Some troubled people avoid all visual stimuli by shutting their eyes for long periods of time. Catatonic schizophrenics, for example, are said to spend prolonged periods of time with their eyes closed, averted, fixed on a particular inanimate object, or just staring out into space.

Although schizophrenics are known generally to avoid mutual gaze more than "normals," it has been found that the topic of a discussion can mediate this behavior. Rutter (1976) showed that adult schizophrenics, who had been described as gazing less at others in general, actually showed this tendency only when the discussion was more personal, i.e., was related to their patient status. Results such as those reported by Rutter (1976) must be taken as a caution to diagnosticians to consider the situational variability of their diagnostic criteria (Fehr & Exline, in press).

Laing (1960) reported that schizophrenics actively fear the gaze of another person and are especially vulnerable to feeling exposed by the looks of others. This fear is found among paranoid schizophrenics who feel eyes "looking at them" and may dread being looked at directly or fear the destructive effect of a glance (Laing, 1960). It is not surprising that the perception of eyes on the Rorschach Ink Blot Test may be diagnostic of paranoia (Hertz, 1961).

Extremely fearful or paranoid people may fantasize that their glance, or that of others, possesses evil destructive powers (i.e., a look that can kill or make others insane). A paranoid delusion may involve a fear of persecution by accusing and condemning eyes. For example, a woman in therapy described her fear of "a shapeless figure looming in the darkness with crazy eyes. No matter where I go in the house it always watches me." In seriously disturbed cases, a person may have a delusion of her/his eyes being mutilated or pulled out (Hart, 1949). Common themes center on changes in the outward appearance of the eyes (e.g., color and form) or on diminished visual ability (Greenacre, 1926).

The following dream, described by a hospitalized person, eloquently illustrates fantasy connected with the eye (the spelling and grammar of the passage have been left unedited):

> I walk through the door and I see a giant blob with one huge eye in the middle. It looked at me and I looked at it. I did not know what to do. It looked at me again. The eye opened like a mouth. It felt like a huge vacume cleaner sucking me closer and closer until we were looking eye to eye. Then he sucked me into its bloby body. Then I just remembered that I had some smoke bommes and matches. So I lit them. Then that blob opened his eyes and couft me out.

The perception of eyes can be not only an indication of severe stress but also of self-consciousness. For example, a woman became self-conscious in a group therapy situation when someone talked to her. She told the group members, "I feel all your eyes upon me." When she was a child, the watching eyes meant judgment and condemnation; she now projected these feelings onto others. This woman is not unusual, as Fritz Perls (1969) has pointed out: "Many people have no eyes. They project the eyes, and the eyes are to quite an extent in the outside world and they always live as if they are being looked at. A person feels that the eyes of the world are upon him. He becomes a mirror-person who always wants to know how he looks to others. He gives up his eyes and asks the world to do his seeing for him. Instead of being critical, he projects the criticism and feels criticized and feels on stage. Self-consciousness is the mildest form of paranoia" (pp. 36–37).

In some cases, the psychotic individual may cherish the gaze and "grab" another with the eyes in prolonged and intense eye contact (Laing, 1960). These people may have weak interpersonal barriers and their stares may reflect a symbiotic desire to become psychologically merged with the other person or to become a part of the other. The person with "devouring" eye contact may yearn for the protection and omnipotence of the other. (Only in the extreme case, however, is staring a symptom of serious problems.)

## DEPRESSION

Depression is characterized by a decreased amount of eye contact (Heaton, 1968). In an experiment using a silent videotape film of five depressed and five nondepressed psychiatric patients, Waxer (1974) noted that depressed individuals maintained eye contact for only about one-fourth of the time of that maintained by nondepressed patients. Waxer (1974) also found that judges who viewed the video-tapes were able to distinguish depressed from nondepressed hospital patients and that visual behavior was an important cue. In another type of study, Hinchliffe, Lancashire, and Roberts (1971) found that depressed people demonstrated a decreased amount of eye contact when compared with individuals who had recovered from depression. These authors suggest that reduced eye contact is part of a defensive withdrawal. Alternatively, low eye contact could be a reflection of the reduced energy state of the depressed individual. In a study on the neurophysiological aspects of depressed states, Whatmore and Ellis (1962) found severity of depression was reflected in a slower speed of eye movement. Their finding correlates with Dunlap's observation (1927) that states of depression or grief include dull, lifeless eyes.

Baker (1967) pointed out that eyes reflect the fear and shame which are reactions to admonitions against looking or staring. Argyle and Cook (1976) noted that depressives avoid mutual gaze because of feelings of guilt and shame. They further hypothesized that the averted gaze of the depressed person is related to low self-esteem. It may also be that a depressed individual avoids eye contact in order to reduce intensity of emotion (Hinchliffe et al., 1971). Or the person may wish to repress or deny emotion and gaze avoidance may aid in this function. This seems to be true in the case of women, who represent the majority of depressed patients (Altman, 1971). The depressed woman may feel angered by something she experiences and also guilty or powerless to express that anger and effectively change the situation. One way to reduce intense emotion and the double-bind of conflicting emotions is to avoid the eye contact which is psychologically arousing.

As in the case of schizophrenics, the possibility of situational variability should be taken into account by diagnosticians and future researchers when relating patterns of eye contact with depression. Concerned with the lack of attention given to the possibility of situational effects on diagnostic criteria, Rutter and O'Brien (1980) compared the visual behavior of withdrawn and aggressive "girls" during discussions of personal and impersonal topics. Similar to the pattern found among schizophrenics (Rutter, 1976), Rutter and O'Brien (1980) found that withdrawn "girls" gazed at their conversation partner less than the aggressive "girls" *only during the personal discussions.* Although this study did not observe depressives per se, the similarities between withdrawn and depressed persons [i.e., clinical descriptions of depressives often include the term "social withdrawal" (Rutter & Stephenson, 1972, p. 64)] suggests that there may be situational variability among depressives as well. At the very least, a need for more research is indicated.


## OTHER PSYCHOLOGICAL STATES

Eye contact can be indicative of other psychological states in addition to autism, schizophrenia, and depression. For example, anxiety is usually discerned when the individual constantly looks down or away from any other, or when her eyes move from one place to another. High blinking rates or fluttering of the eyes may also indicate nervousness or discomfort. The individual's eyeblink rate may tend to increase as the amount of perceived fear or threat increases (Appel, McCarron, & Manning, 1968). Under secure conditions, eye movement is slow and extensive. When normal subjects are asked an anxiety-producing question, the extent of eye movement is greatly reduced and the velocity greatly increased (Day, 1967).

Persons who have experienced series of personal losses during formative years (for example, those forced to live in foster homes and/or with inconsistent human attachments) display characteristic patterns of eye contact. "In the psychotherapeutic situation these individuals rarely seem able to sustain eye contact for more than a fleeting moment. . . . One gets the impression that they have never laid eyes on a contented face, nor do they feel capable of evoking a smile or friendly nod from anyone" (Ratnavale, 1984, p. 8).

The very complicated syndrome of multiple personality, an extreme form of dissociative hysteria, can be characterized by changes in the eyes. When a person enters an amnesic fugue state, a strange look often overtakes her/his eyes. The following description is from a woman who was diagnosed as a multiple personality. She wrote as Sue, the main personality who remembers the others. Her report illustrates how eyes and personality are interrelated, even to the extent of physical disturbance in the eyes resulting from psychological problems.

> I, as Sue, usually find it comfortable and pleasing to make eye contact. My family tells me that my eyes become "crystalized" when I get nervous. They say that during this glassy look a noticeable change in personality takes place. They describe each personality as having a totally different look as far as the eyes are concerned. For example, Mary (who has not cried since age 3) is described as having the eyes of a frightened fawn, wide-eyed and skitterish as though anticipating danger but too young to protect itself.

My family describes Jean's eyes as being fiercely intense . . . That the person in the path of their glare often feels as though Jean's eyes could burn holes through them. Jean usually makes others look away from her because it becomes too uncomfortable for them. I can't look at her eyes in the mirror for any length of time. They appear evil and frighten me. Jean wears reading glasses, Sue does not.

Mary . . . likes eye contact but becomes easily embarrassed, will look away for a moment, but usually returns to eye contact soon. Mary's mind wanders frequently and my family says that when that happens she will continue to speak and look at you but her eyes appear to look right through you.

Janet has an honest look in her eyes and her eyes appear at the same time to be both innocent yet very wise. She likes eye contact and it's easily established and held. My family knows Janet even before she begins to speak by looking at her eyes. My sister says that Janet's eyes sparkle and bubble and my brother says that you expect them to snap, crackle and pop. Janet's eyesight is extremely keen, especially at night.

The Young Mary . . . established eye contact only when frightened, which is most of the time. Her eyes are wide and big and she seldom blinks. My mother describes Mary's eyes as very fluid, as though floating in a bucket of tears, yet no tears ever fall. Mary establishes eye contact for the sole purpose of anticipating that person's next move. Once she trusts you and is no longer afraid of you, she will rarely raise her eyes.

Sue's sharing of her experience provides special evidence for clinicians. Her fascinating account illustrates quite clearly how personalities can use eye contact patterns and even differences in eyesight as unique expressions of their individuality, and how such individual differences can be easily identified by others. In Sue's case, one human body exhibited a variety of gaze patterns and visual capacities. Her experience suggests that personality characteristics may be of greater importance in determining eye behavior and appearance than genetics or cultural influences.

Tension and stress, which often involve eye, neck, or head muscles, may lead to complaints involving visual disturbances (Clancy, 1963; Greenacre, 1926). Ophthalmologists have claimed that psychosomatic eye symptoms are extremely common (Schlaegel & Hoyt, 1957). They have also reported on the emotional etiology of eye disease (Inman, 1935, 1955). A high percentage of depressed people complain of eye trouble (Birge, 1945; Grinker, 1961) and nearsightedness and astigmatism often exist in cases of schizophrenia (Reich, 1972). Clancy (1963) suggested that eye defects could be the vehicle for expressing emotional conflicts.

It should be stressed that eye contact is but one of many behaviors, verbal and nonverbal, which the skilled clinician notes in her/his assessment of a client. Eye contact may provide clues for diagnosis because a person's eye contact and general looking patterns reveal much about her or his personality and emotional state. Severely disturbed individuals may make little or no eye contact at all, such as autistic children, or they may make prolonged and intense eye contact, which is found in those who are overly dependent or needy. In less acute cases, abnormal gaze patterns may result from anxiety or from some other situational condition (e.g., physical problems). The diagnostician's observation of this nonverbal behavior can be of valuable assistance in assessing the nature and extent of the client's problems.

The diagnostic value of eye contact can also aid the psychotherapist in charting the progress of the client by gauging not only the immediate emotional and mental state of the individual, but also changes in the client's patterns. The therapist can notice subtle changes in the client's eye contact and can judge the appropriateness of the client's eye behavior to the situation. For example, the person who stares into the therapist's eyes may be expressing a dependency on the therapist or even a wish to adopt the therapist's identity. A client may feel, for instance, that if s/he would merge with the therapist's personality then s/he would no longer have any problems. From this behavior of persistent staring, a decrease in the amount of eye contact may indicate more comfort with her or himself and more a sense of her/his own identity.

More commonly, though, when people avoid eye contact, they are avoiding personal interaction. The lack of mutual gaze with the therapist, who is initiating it, may reflect the client's fear of self-knowledge and awareness. Greenson (1967) discussed how the avoidance of eye contact as the client enters or leaves the therapist's office is an indication of resistance. An increase in mutual gaze with the therapist, however, may indicate a desire to make contact with another person and may accompany a greater ease in dealing with others in the client's daily life. In general, eye contact on the part of the client probably indicates comfort, affection, and lack of resistance (Greenson, 1967; Pfeiffer, 1968).

As a person becomes ready to accept her/his feelings (no matter what they are), eye contact may change dramatically. A sudden change, then, may signal positive progress of the client. It may even be fostered by the therapist. An illustrative case involved a child who was unable to look at the therapist. Session after session, the child sat on the sofa with his gaze averted, trying to understand the source of his problems. In order to have some sense of contact with him, the therapist started sitting on the floor in front of the child. Finally, one day the child talked about his anger toward his mother; as he began to accept this feeling, he gave up his sense of shame, and for the first time in therapy, the child established eye contact with the therapist. Thereafter, he displayed no similar symptoms.

The gaze of the therapist is also a significant aspect of the clinical relationship. The look of the therapist conveys attitudes to the client and may be one of the more important curative elements in psychotherapy. The expressions of "a lifted eyebrow, a look of shock, or scorn or dismay, a vacant stare . . . tell the clients much more about our feelings toward him than our words do," asserted Hammond, Hepworth, and Smith (1977, p. 184). They pointed out that, "if a discrepancy exists between the counselors' verbal and nonverbal communication, the client is more likely to attach credence to the latter than the former" (p. 185).

## EYE CONTACT AND THERAPY

Some individuals, whose lives are very troubled or barren, make their first truly human contact in months or years within the therapeutic relationship. Moments of eye contact in therapy can be times of deep communication and closeness, and the optimum use of eye contact between therapist and client facilitates the counseling process (Brown & Parks, 1972). The eyes symbolized "healing" as early as 3000 B.C. (Martí-Ibañez, 1962). They are superb vehicles for expressing warmth, acceptance, caring, empathy, respect, and interest.

Although verbal communication is usually predominant in traditional psychotherapy, recent studies have emphasized the significance of nonverbal communication between therapist and client (Haase & Tepper, 1972; Hamm & Hartsfield, 1970; Hodge, 1971). Facial expressions, as well as other body and nonverbal cues, may communicate more about the interpersonal attitudes of both the client and the therapist, and have more of an impact, than do verbal messages (Argyle, Salter, Nicholson, Williams, & Burgess, 1970; Hammond, Hepworth, & Smith, 1977). A change in nonverbal cues has more than four times the effect as a change in verbal cues (Argyle et al., 1970). Eye contact in particular is a means of conveying attention, which "allows one to receive verbal and nonverbal messages, conveys respect and interest, and serves as a powerful reinforcer" (Hammond, Hepworth, & Smith, 1977, p. 110). Thus, it is recommended that the therapist be careful to maintain a frequent level of eye contact and to lean towards the client while expressing her or himself. Such nonverbal cues of immediacy have been shown to improve significantly observers' ratings of a therapist's interpersonal skills and effectiveness (Sherer & Rogers, 1980).

The kinds and qualities of eye contact and eye behavior experienced by the clinician are as many and as varied as the clients themselves. It is indeed a strange experience for a therapist to be seated across from an individual who is not looking at all or from one who is staring intently, almost "consuming" the therapist's eyes. It is equally unnerving for the clinician to try to relate to someone who stares "through" her, such as an autistic child who behaves as if the therapist does not exist. Within these extremes, of course, exists a wide range of more common behavior to be observed and interpreted.

Often the clinician may not be aware of the role that eye contact is playing in the therapeutic interaction until the eye contact overtly changes. In one case, a man was extremely shy, nervous and embarrassed, and unable to touch people (except during sexual encounters). He usually exhibited a great deal of nervous movement, with his shoulders suddenly jutting out in space or his body tilting off its axis; an arm might spontaneously reach across his body and grip the other arm. At first the therapist did not notice that the client was not looking at her. Then in one session he began to look into his therapist's eyes, and both felt a type of closeness they had never experienced with each other, even though they had held hands or hugged on previous occasions. The client said that this mutual gaze was the most intimate kind of contact that he had ever made and meant more to him than touching. He said he felt less "scattered" and more "centered" as he looked into the therapist's eyes. At the same time, his body became more calm.

Usually, however, changes in eye contact are gradual and subtle. For example, a client who had had an emotional "breakdown" began to feel more positive about herself. During the breakdown, she was unable to make eye contact with anyone; however, as she felt better she resumed mutual gaze. The members of her therapy group were astonished at the difference in her ability to make eye contact and at how much more comfortable they felt with her. Her direct gaze then became, as with many people, a "barometer" of her feelings of self-confidence and self-worth. Clearly, the therapist's observation of the client's emotional state, as revealed by eye contact (or lack of), contributes significantly to understanding the client.

Eye contact is often important in establishing the client/therapist relationship. As one client remarked: "The first thing I noticed about my therapist was her warm eyes." When therapist and client first meet, the therapist seeks to establish a relationship through which certain therapeutic attitudes may be communicated. Eye contact, more than other nonverbal behaviors, lets the client know that the therapist is attending to her/him with caring, warmth, and interest (Sobelman, 1973; Hammond, Hepworth, & Smith, 1977).

On the other hand, the therapist's lack of good eye contact could result in a negative response from the client, such as this client's emphatic description of a therapist: "He couldn't even look at me directly. I felt he was afraid of me" (Agel, 1973, p. 12). Poor or minimal eye contact by the therapist may even preclude the establishment of a relationship with the client at all. Notetaking may become an excuse to avoid contact with the client. In one case, a woman decided very soon into the initial interview that she found the therapist's behavior unacceptable for her needs:

> I entered his office and sat down. He sat down behind his desk and immediately pulled a yellow legal pad towards him. Dr. X readied his pen with a loud click, lowered his head and asked, "Now, what seems to be the problem?" I gulped and uhmmmmed, and *he* started *writing*. I gathered my thoughts, much as I would before speaking into a tape recorder, and then began to describe the reasons I wanted therapy. He "uh-huhed" and nodded throughout my speech, all the while writing with his head down. At one point, I paused and his hand paused, just like that, his lowered head suspended on a nod. I continued, putting him back in motion. When I finished he looked up, met my eyes briefly, then glanced down at his notes.
>
> I felt he regarded me as a clinical entity and what I wanted was human contact. I resented his dispassionate approach and wouldn't be back to see *this* therapist.
>
> I then decided to find a feminist psychotherapist and did so. This woman helped me to feel comfortable and to explore my own problems without dehumanizing me. One essential element in our relationship was eye contact. The warmth of her eyes encouraged me to accept myself in a way I never could have with a therapist who kept his head down, writing notes.

It would be interesting to assess the different qualities and patterns of eye contact among male and female therapists with both male and female clients. Do female therapists have more eye contact with their clients than male therapists, since this pattern is typical of men and women in general? Do female therapists sit at an angle from their clients which is more conducive to eye contact? Do some male therapists stare for long periods of time at their female clients?

The ability and personal orientations of the therapist also influence eye contact in a therapeutic interaction. The therapist who is comfortable with intimacy in her or his own life will probably have more eye contact with her/his client than the one who is avoiding such interaction, and the therapist who fears contact with people will most likely avert gaze as a means of keeping distance between her/himself and others. In addition, the therapist's prejudices may very well be evidenced by decreased eye contact. The client, in turn, is likely to respond by feeling discounted and misunderstood.

Bigotry can be communicated by looking less or by the gaze behavior of dominance and status. The therapist who "stares down" a client may serve to intimidate the client and increase her or his feelings of low self-esteem. Thus, the eye contact behavior of the therapist enables a client to discern the therapist's attitudes toward different groups of people and the therapist's potential for effectiveness in working in relation to these people. For the individual who is seeking a therapist, eye contact during the initial interview is a helpful indicator of the potential rapport between them. The choice to continue a therapeutic relationship is often based on the "feel" of the first interaction and much of this feeling is promoted by eye behavior.

Not much data exist on how people react to one another in an initial interview. One study by Kleinke, Staneski, and Berger (1975) showed that nongazing interviewers were rated as least attractive, were given the shortest answers, and that subjects sat farthest from them during a debriefing session. In another study (Kelly, 1972), subjects judged most highly the photographs of therapists who sat close to clients, face-to-face, and maintained eye contact. Furthermore, a greater use of mutual gaze tends to enhance the client's regard for the therapist (Villa-Loroz, 1975). Indeed, for some clients, eye contact is a required component of therapeutic relationship. One reported:

> I saw a female therapist who had a psychoanalytic orientation except that she did not employ a couch. I noticed that, when I expressed an emotion, she looked away from me. When I questioned her about her behavior she explained that she looked away at emotional moments so as to maintain her objectivity. I told her I found this intolerable; I left the session knowing I would not continue consulting with her. Eye contact had become a therapeutic necessity for me. . . .

The seating arrangement of therapist and client naturally plays a large role in the amount of possible mutual gaze. In the usual psychotherapeutic setting, the client is seated across from the therapist at a forty-five to ninety degree angle. Since some clients become uncomfortable with the face-to-face encounter, this angle gives the client the option of looking at the therapist or avoiding the therapist's visual scrutiny. Sullivan (1954) felt that the right angle position was ideal for psychotherapy, as well as mandatory with schizophrenics because of their fear of being watched. In this position, the therapist may not always be able to perceive the fine facial expressions, but s/he can use peripheral vision to obtain most information. However, many therapists do consider direct eye contact to be essential in the psychotherapeutic relationship. For instance, Virginia Satir (1976), a highly skilled and well-known family therapist, stressed the importance of maintaining eye contact when there are vast differences in height among family members. She recommended equalizing heights (using stools, chairs, etc.) when necessary to facilitate eye-level contact.

Direct eye contact, however, is not considered to be of value in some forms of therapy. In traditional analysis, a prone position with the analyst behind the couch precludes eye contact. Freud's awareness of the influential power of eye contact contributed to his belief in the "couch" method. He did not want his facial expressions to give the client material for interpretation or to influence the flow of unconscious thoughts which may become verbalized (Ratnavale, 1984). If the client is seated in psychoanalysis, most commonly s/he is so far from the therapist that direct eye contact is difficult.

Although psychoanalytic techniques can facilitate self-awareness, the analyst should guard against so rigid a use of the couch that it prevents contact. An analytic client described this necessity well:

> One day in psychoanalysis I felt the need to look at my analyst, so I sat up across from him and looked deeply into his eyes. After months of on-the-couch psychotherapy, this was the first and most intense experience I ever had with him. I felt a closeness which was indescribable. In this encounter, I experienced caring, affection and humanness in the warmth of his eyes. I felt that it was a time in which my therapist was open to me and had dropped the guarded analytic posture. I tried to recreate the experience with him. I wanted so desperately to reach into his eyes one more time, but it never happened.

In most therapeutic approaches, therapist and client sit fairly close, more or less face-to-face. Scheflen (1963) noted some details of the therapist's looking behavior with a client during a session. When the therapist was listening, her/his head was slightly lowered and cocked to one side so that her/his eyes were averted. While commenting or interpreting, the therapist might raise her/his head, look directly at the client and speak. When finished, the therapist often looked away.

## THERAPEUTIC ATTITUDES

The therapist's behavior, both verbal and nonverbal, serves to communicate therapeutic attitudes to the client. Some of these behaviors and attitudes will facilitate positive change in the client. Further investigation is needed to specify the exact behaviors and characteristics relevant to the therapeutic change, as well as to define the role of each in the total communication process. This need has been pointed out by Truax and Wargo (1966), who discussed the importance of nonverbal aspects—including physical distance (Kleinke, 1972), voice quality (Markel, 1969), head and body position (Mehrabian, 1968a), touch (Jourard & Friedman, 1970), facial variables (Ortega y Gasset, 1957), and eye contact—in communicating therapeutic attitudes.

When therapeutic change occurs, one or more of these behaviors usually plays a significant role.

Carl Rogers (1957) identified three attitudes which are characteristic of an intimate relationship and are conducive to positive therapeutic change. Each of these attitudes—empathy, positive regard, and congruence—is fostered by eye contact. A study by Webbink (1974; see also Chapter IV) substantiated this statement. The subjects in the experiment reported feeling heightened empathy, greater willingness to reveal themselves, and more positive feelings toward a confederate of the same sex after a brief experience of mutual gaze than they did in two comparable nonverbal situations. Since empathy, positive regard, and congruence are crucial elements in most therapeutic relationships, each of these attitudes and its relevance to eye contact will be discussed.

Empathy is the sensitive ability of a person to see the world through another's perspective. It is a warm acceptance of the other person, based on appreciation and respect. In a study by Marks (1971), people who maintained eye contact with the experimenter had higher scores on an empathy scale than those who avoided eye contact. Marks also found that empathy was felt to be stronger when mutual gaze occurred.

Empathy is the attitude most related to positive therapeutic change (Barrett-Lennard, 1959) and eye contact is the most significant variable related to the perception of empathy (Katz, 1965; Marks, 1971). Haase and Tepper (1972) investigated empathy ratings of interaction segments which were shown on videotape to professional counselors. They found that those interactions judged as the optimum communication of empathy were characterized by eye contact in combination with (a) forward lean of torso, medium empathic verbal message, and a far distance, or (b) forward lean of the torso, high empathic verbal message, and a close distance. The interactions that seemed to have the least communication involved no eye contact.

Rogers' second therapeutic attitude, positive regard or feeling, is the most widely researched component of eye contact. In several studies, Mehrabian (1967, 1968a, 1969a) either instructed subjects to assume a physical posture toward someone they liked, or asked them to rate photographs of people (who were in various body positions) for likeability. Focusing on the frequency of eye contact as a component of each likeable assessment, he reported that eye contact and likeability are related. Kleck and Neussle (1968) also observed that the amount of eye contact in a relationship appears to vary directly with positiveness of affect and with the degree of attraction. It has been shown that a higher percentage of eye contact reflects a more positive attitude between communicators (Kleinke, 1972; Mehrabian, 1967, 1968a, 1969a; Mehrabian & Friar, 1969). The positive feeling between therapist and client may contribute to a relationship which is warm, open, and caring.

The willingness to engage in mutual gaze is greater in persons who are judged high in the desire to establish warm interpersonal relationships (Exline & Winters, 1965a), and this is characteristic of many therapists. Some clients, however, might be judged low in this desire because of problems with poor self-esteem, depression, anxiety, etc. The therapist's positive regard may serve to increase not only the client's feeling of self-worth, but concomitantly increase the client's desire to engage in close interpersonal relationships. In a study similar to that of Haase and Tepper (1972), Kelly (1972) had subjects rate photographs of therapists for their communication of "liking." Eye contact was found to be a major component of a high rating.

The positive regard of the therapist may also encourage greater eye contact by the client. Researchers have determined that, when a subject is confronted with an approving and a nonapproving confederate, s/he will look more often at the approving one (Efran, 1968; Pellegrini, Hicks, & Gordon, 1970). It is not always the case, however, that the therapist's eye contact is perceived positively by the client. In a situation that is nonapproving, such as when the client's negative aspects are discussed, the therapist's gaze may be unwelcomed and disliked. Ellsworth and Carlsmith (1968) found that, in a conversation indirectly critical of the subject, the more a male interviewer looked, the less the subject liked him. It has been suggested that the therapist make eye contact only when the client is making positive self-statements (Brown & Parks, 1972).

Congruence, a third condition which Rogers (1957) identified as indicative of an intimate therapeutic relationship, occurs when the feelings within a person and the outward expression of these feelings are in harmony. In other words, one's experience and awareness of self are integrated. Instead of presenting an outward facade, the congruent individual is sincere (Kleinke et al., 1973). People are highly congruent during peak moments of emotional and physical expression, such as when experiencing grief, anger, fear, and sexual passion. For instance, a person is congruent when angry feelings, boiling inside, are expressed in flashing eyes, direct gaze, aggressive forward-jutting jaw, loud screams, and clenched fists.

Although eye contact can be controlled, the eyes may reveal more than a person chooses. For this reason, incongruence is often manifested as a contradiction between the expression of the eyes and that of other parts of the face or body. For example, a woman who was asked by a family member to attend a therapy consultation denied she had any problems. Although she successfully controlled the rest of her face, her agony was apparent in her eyes, which were glossed by a film of tears. Sadness and/or depression are often recognizable only through the eyes. While an individual may be communicating both verbally and nonverbally a sense of happiness and well-being, her eyes may show sadness. A change in eye expression, especially teariness, often indicates to the therapist that it is appropriate to ask about the client's feelings or to comment on the sadness. If the client does not deny distress, this perceptive statement by the therapist may well facilitate the client's congruence through a greater intensity of emotion. An awareness of one's true feelings or the "tears behind the eyes" (Berryman, 1971) is a mark of therapeutic progress. (Of course, the clinician must be careful to check other reasons for moisture in the eyes, e.g., contact lenses or allergies.)

## THERAPISTS AND TECHNIQUES

Sometimes the silent eye contact between client and therapist, between one client and another, or between a client and her own eyes in a mirror will result in the client's discovery and expression of feelings in an integrated way. One client reported:

> I thought I was angry at my therapist. I felt defiant, feisty, coiling my body in the chair opposite her, ready to attack. She must have sensed very different feelings coming through my eyes, so she handed me a mirror. I looked into my own eyes and felt fear. I stared at myself until the fear flowed through my eyes to my forehead and mouth. It grew into a scream that filled my whole body until the tension was gone; my eyes grew softer; I relaxed; felt whole again, in touch with myself. (Malone, 1975)

Eye contact can intensify emotion whether it is positive or negative (Ellsworth & Carlsmith, 1968). When people gaze into one another's eyes for a long time, feelings are communicated which are difficult to verbalize. Individuals often feel relieved when the emotions they have longed to share, but were frightened to reveal, have been expressed. The therapeutic atmosphere should be safe and supportive, and the therapist's eye contact reassuring. An example of this intimate eye contact in therapy occurred with a woman who had major problems with closeness and trust. She described her marriage as emotionally sterile, and came to therapy for love and understanding, as well as for greater awareness of herself. The profoundly moving eye contact happened one day between her and the therapist, as they looked silently into one another's eyes for a long time. Tears streamed down the client's face as she felt the caring, warmth, and empathy of the therapist. The client said that this interaction was the most meaningful experience she had had in her life.

Many people have commented on the special intimacy of the therapeutic situation. Eye contact can be one of the most intense aids to this intimacy, such as in a case described by Seagull (1967): "We continued to look wordlessly at each other . . . We sat in complete silence for a while feeling very much together and at one with each other" (p. 43). Although the exact behaviors and characteristics relevant to change are not strictly determined, communication of three Rogerian therapeutic attitudes—empathy, positive regard, and congruence—clearly assists the client. Mutual gaze is a significant nonverbal behavior in communicating these attitudes, as well as in establishing and maintaining a caring, intimate relationship between therapist and client.

Eye contact can also be a technique used to penetrate the individual's emotional barriers. Therefore, the therapist should often react to an averted gaze by encouraging the client to look at her/him. This encouragement can be particularly effective when the client is expressing anger toward the therapist. In one such case, when a woman was timidly expressing anger, the therapist asked that she make eye contact. The woman looked at her therapist, her eyes filled with tears, and she said, "I can't; I can't tell you." Then, as she became uncomfortable with the intimate gaze and lowered her head, the therapist reached over and held her. The woman cried for the first time in months of psychotherapy.

To cite another example, a client who was usually quite defensive and looked down during most of a session tentatively began to express anger toward the therapist. This man often had negative feelings toward other people, which he continually described in his therapy sessions. However, he would never discuss his feelings with the person concerned. This was his way of maintaining power in relationships, a value which he held very high. He had some deep-seated anger stemming from his early childhood because his parents were working-class and of immigrant background. He felt humiliated by their position in society and, as he grew older, he attempted to deny this heritage completely, by leading an upper-class lifestyle. The anger and secretiveness that he nurtured alienated him from others, and he was without friends.

As the therapist encouraged him to express his negative feelings directly, he began to stumble over his words and look away. The therapist, however, insisted he look into her eyes, causing a novel situation which changed the whole atmosphere of the session.

Bravely, for the first time, he expressed his anger directly and, instead of letting these feelings fester inside of him, he found that he felt close to another person. After a long period of silence, the therapist reached over to him and said, "I care about you." He paused and responded, with tears in his eyes, "I wish I believed you." As he silently gazed into the therapist's eyes for a long time, a change seemed to take place. He saw caring in her eyes and he began to realize that not all people were "after him." He was now able to trust and, at the same time, to accept the anger that he had always nourished as a dark secret.

People frequently look away while expressing feelings which are painful or difficult for them. Since looking away often accompanies feelings of shame and discomfort, progress is made when the client begins to confront feelings and to look directly at the therapist while expressing them. Eye contact may then occur spontaneously in the client who begins to realize potential, and s/he may experience a new sense of openness, spontaneity, and honesty. It is evident, therefore, that eye contact can be used by the therapist to help the client confront her or himself.

In these examples, eye contact served to underline feelings. Eye contact can also intensify both positive and negative feelings. If, for instance, one says, "I love you" or "I hate you," the lack of eye contact may detract from the expression. Alternatively, its presence may increase the intensity of a statement. The therapist who wishes to emphasize an interpretation of a client's situation may say it while focusing on the client's eyes, thus reinforcing the client's attention to this statement. In therapy, then, eye contact is both a positive and negative reinforcer (Brown & Parks, 1972).

Eye contact can be of central importance in behavior therapy, both in the establishment of a relationship with the therapist so that other behaviors may be modified, and as a behavior which changes as a result of the reinforcement of other behaviors. Behavior therapists quite literally choose eye contact as one of the first behaviors to condition because it can be so easily reinforced (Bate, 1972; Etzel & Gewirtz, 1967; Page, 1971; Sieveking, 1972; see also Chapter II). This choice reflects the therapist's need for the focused attention and emotional relatedness of the client, especially one who has difficulty attending to the therapist.

Many treatment studies of autistic children first condition eye contact (Irwin, 1971; McConnell, 1967; Maurer, 1969; Ney, 1973; Saposnek, 1972). Even with these difficult children, researchers have found that eye contact does increase with reinforcement. Ney (1973) studied a 4½-year-old autistic boy and found a significant increase in the frequency of mutual gaze with both *contingent* and *noncontingent* reinforcement. In one of the earliest studies of its kind, McConnell (1967) treated a 5½-year-old autistic boy for gaze aversion. He reinforced gaze by smiling and praise, and also repeated the praise when the boy made eye contact after an intelligible utterance. The number of gazes in a 45-minute session rose from 13 to 180 within 33 sessions. Similarly, Colligan and Bellamy (1968) found that their efforts to increase the length and frequency of eye contact with autistic children were relatively successful in a study in which they played peek-a-boo using a mirror, cloth, and large pair of glasses.

In general, behavior modification techniques have helped hostile and withdrawn children establish meaningful relationships (Exline, 1971). In 1969, Maurer also used the peek-a-boo game with a withdrawn child and because of his success, he hypothesized that the game might be useful for therapy. In a more recent study, the behaviors characteristic of childhood depression (including lack of eye contact) have been effectively treated by such behavioral techniques as instructions, modeling, role-playing, and feedback (Frame, Matson, Sonis, Fralkov, & Kazdin, 1982). In the case of schizophrenic adults, Wallach, Wallach, and Yessin (1960) reported that encouraging the schizophrenic to look into the therapist's eyes helps decrease rapid eye movements, as well as increase thinking based on reality percepts. When rapport has been established, eye contact is divested of anxiety because a sympathetic, nonanxiety-provoking response occurs in its place (Wolpe, 1958).

In addition to the behavior modification approach, eye contact is facilitated by other techniques. In such therapies as bioenergetics, Rolfing, Lomi work, and Reichian therapy, the client is encouraged to express feelings through the body. Using one of these techniques, Reichian therapy, the therapist may focus on the muscular tension around the eyes by deeply massaging this area to the point of pain, ultimately releasing repressed feelings. This release may help the therapist and client to look at other feelings which may have been repressed or inhibited. For instance, if the client saw something fearful or traumatic early in life, s/he may now squint her eyes in an attempt to reduce visual stimulation. The client may feel unconsciously that if her eyes were opened, the feared scene would be reexperienced.

In an interesting case, a primal therapy client had a form of severe myopia, as well as an inability to focus intently on others, and was unable to relate to people through his eyes. With intensive psychotherapy, he found that not only did his visual problems dissipate, but also that he could no longer tolerate his relationship with his wife. He realized that he had unconsciously married a woman similar to his mother so that he could work through the ambivalence he felt toward her. He recalled that his mother, due to her intense fear of intimacy, constantly avoided eye contact. Her eyes would dart around the room, never staying fixed on another person's eyes. It was no accident that this man married a woman with the same kind of eye pattern. Interestingly, he noted that his wife's father was a very seductive man who would look at her for long periods of time. She would return his gaze for only a brief moment, then look away. The client and his wife had two daughters with the oldest child modeling herself after the father and the youngest after the mother (she too had darting eyes). After divorcing his first wife, the client married a woman who maintained intense eye contact. The child who maintained more eye contact decided to live with her father, whereas the younger one stayed with her mother.

According to Reichian therapists (Konia, 1970; Nelson, 1976), the contraction and immobility of the ocular segment (or "eye block") restricts emotional expression and contact with reality. Konia (1977) claimed that this ocular armor is the primary block in schizophrenia; however, it is also found in less severely distressed cases. In Reichian or orgone therapy, the eye muscles are directly stimulated; sometimes the client opens the eyes wide, then rolls the eyes and may scream (Reich, 1972). This action is significant in liberating the ocular armor. The therapist establishes eye contact with the client by encouraging the expression of feelings through the eyes and by attempting to dissolve the ocular armor (Blasband, 1967).

There are several other types of therapy which employ the eyes in facilitating awareness and catharsis. Eye contact is used in Gestalt therapy and psychodrama when the client tries to recover repressed incidents through roleplay. In the psychodrama, the client may make eye contact with other actors and thereby become more emotionally involved. In Gestalt therapy, a person may be called upon to speak to someone who is not actually there. When asked to imagine looking into the other's eyes, the person may express intense feelings. In many cases, the person cries and begins to feel "present" in the situation. The very same Gestalt exercise may prove far less intense without the instruction to make eye contact, especially if the imagined person is a loved one—perhaps someone who has died. For example, a woman in therapy was unable to express her feelings about her dead cat. When asked to talk to the cat in fantasy, she imagined grasping the cat's front paws and staring intently into the cat's eyes. "I am witnessing the moment of death and seeing your life leave your body through your eyes. I see the sparkle and the aliveness just vanish."

Similarly, a woman who had frequent unexplained crying spells was asked to pretend she was talking to her father on his deathbed. She sobbed as she said, "I look into your eyes, you who have always been so strong and vibrant. Now you cannot move or speak. Your eyes cry out to me to help and I can do nothing; I love you Daddy. Please don't die." In doing this exercise, she realized that she had been burdened with this horrible memory, and much of the sadness that she had "carried around" was directly related to this incident. After she shared it with the group members, she felt as if a weight had been lifted from her.

Using another approach, Guided Imagery and Music (Bonny & Savary, 1973), a client is blindfolded and a deep state of relaxation is induced. While listening to music, one client reported feeling with no display of affect. It became clear that he felt it unmasculine to express deep feelings, especially affection for his son. When he talked about his son, he said he did not like children and was very glad his ex-wife had custody of them. At some point during his fantasy, however, he had an image of being alone with his son. The therapist used a Gestalt technique and asked him to imagine looking into his son's eyes. For the first time in many years the man began to cry. The awareness of his depth of feeling enabled him to decide eventually to take custody of his children.

Approaches such as Gestalt therapy, bioenergetics, psychodrama, Guided Imagery and Music, and Reichian therapy serve to intensify feeling, in much the same way as eye contact. Therefore, eye contact when actual or imagined used simultaneously with other powerful techniques can be a very effective tool. For instance, if a person is analyzing a dream through the psychoanalytic technique of free association, s/he can intellectualize about feelings more easily than if asked to imagine becoming part of the dream or to look into the eyes of the people in the dream.

Eye contact between the client and therapist is crucial to the deep, regressive work involved in "re-parenting" (Ballentine, 1984). In the re-parenting process, the therapist (or group therapy member) cradles the client in her/his arms and maintains as much eye contact as possible. The early emotional experiences of the client with her/his parents are brought back through this recreation of parent–child intimacy. As this work proceeds and the client begins to connect emotions with early experiences with the client's parents, the therapist attempts to provide a corrective emotional experience by repetitively verbalizing caring messages such as "you are very special; you are a sweet, lovable, little boy or girl; you are very beautiful, etc." By accepting these messages more and more fully during the recreated nurturing scene of parent–child mutual gaze and physical closeness, the patient develops a much enhanced sense of self-worth and personal lovableness. Re-parenting is especially suited for people who have suffered from serious early emotional deprivation or neglect (Ballentine, 1984).

Discussing the client's eye contact behavior in the session can also be an effective therapeutic tool. Although nonverbal communication is not usually discussed in the psychotherapy session, Reimer (1955) has urged the therapist to do so in the case of looking behavior. Eye contact is a cogent example of the client's style of relating to other people. Thus, it can be utilized to encourage self-awareness, since it offers a concrete example of nonverbal behavior and is one that can be identified and dealt with directly.

In a unique case, a client consulted with a therapist because of a problem with eye contact itself. He reported that he was an excellent public speaker, but his avoidance of mutual gaze detracted from his ability to make contact with the audience and was a detriment to his chosen career. His wife pointed out that public speaking was not the only area affected; his friends were alienated by his lack of eye contact with them. (Interestingly enough, he did maintain good eye contact with his wife.) The therapist reported:

> I asked Mr. X how he felt his life might change if he made eye contact with other people. He was not sure . . . I noticed that he did not refer to his wife (who was present at the session) by name, but called her "she" or "her." I commented that he seemed to objectify other people and lacked a personal connection with them. At a later point in the session he said that he was bothered by the barriers between himself and other people and that his eye contact problem might stem from a fear of intimacy. He reported that he felt that eye contact was a personal invasion and that he would be on too intimate a level with people if he looked at them. He said that a friend commented, "You won't let me know you." He also noted that when he made jokes or when he was with people with whom he felt comfortable, he would look into their eyes; or if he found a friendly face when he was talking, he would make so much eye contact that the person might feel uncomfortable.
>
> *Treatment Plan:* I gave Mr. X a homework assignment for the next two weeks to carry a notebook around with him and make notes as to how he is reacting to different people as he forces himself to make eye contact with them. He is to notice how their response to him affects him and how the world seems different to him while looking at others. Mrs. X said she would watch him to make sure that he was really making eye contact because often he thought he was making prolonged eye contact when she perceived that he was engaged in a swift glance at the other person. Mr. X reported that he was very observant of other people but that he looked at them only when they were not looking at him. In the next session we planned to review his notes and to delve more into the causes for his lack of eye contact.

In fact, Mr. X cancelled his next session. After only one consultation, he found that, with conscious practice and diligent observation, he was able to be aware of others' reactions to him and to concentrate on making eye contact with people. He was able to overcome his problem by keeping a diary, noting each time he looked at people and when he looked away. This process served as a constant reminder to initiate eye contact. Since his behavior was changing, Mr. X did not feel it necessary to delve into his unconscious motivation.

Another case of an eye contact problem involved a woman with glaucoma. She related a detailed history of embarrassment about visual problems and deformities which led to problems with eye contact. She was also told at one point that her glaucoma would eventually blind her. It was fear and the resultant stress which eventually brought her to seek psychotherapeutic help. As this woman unburdened herself in therapy, she became better able to share her pain with others and to avoid feelings of isolation and loneliness.

> On a gray and blustery February day, Dr. C informed me that I definitely had glaucoma and prescribed adrenalin eye drops for me . . . Without treatment, she said, I could expect to go blind in five years; so I began using the eye drops nightly according to her instructions.
>
> I had given up on contact lenses shortly after Dr. C prescribed the eye drops for me, and gone back to using eyeglasses as I had done without incident for over fifteen years. Now, however, I found myself dropping my glasses at the slightest opportunity, and the closer I observed myself doing this, the more it seemed that I was somehow compelling my arm to throw them on the floor away from me, in spite of myself. My therapist and I decided that as my mental outlook improved, I was striving for the first time in my life towards maintaining a higher level of eye contact with people and that I was actually throwing my glasses away because they came between the world and me. I solved this problem in a rational fashion by not wearing my glasses unless I absolutely needed them for reading or driving. As I progressed in this latest adventure of using my eyes, I somehow reached a new plateau of experience. It took me several weeks to adjust to the impact of the additional sensory input I was receiving. The eye contact with relatives, friends, colleagues, and strangers was very moving for me.
>
> Not long after I made this change in my looking behavior, Dr. C advised me that the eye drops had succeeded marvelously well and quickly . . . She said that it was not likely that my eye pressure would rise again but said that treatment would be available if necessary.

A member of this woman's therapy group provided an eloquent testimony of her progress in opening up to the intimacy of eye contact:

> While looking at others in the group, my attention came to rest on D. It was as if there was a magnetic pull so that our eyes opened simultaneously and focused

on each other. It seemed that the sun came out; there was an incredible sense of presence and immediacy, yet a timeless quality to this touching through the eyes. There was a sense of total clarity, nothing preventing her from knowing her completeness and sensing me as I am. It was one of the strongest experiences I have ever known of being fully in contact with another person, yet retaining a sense of integrity and autonomy from each other as separate, whole persons.

Before closing discussion on the use of eye contact in individual therapy, a word of caution must be added. As mentioned throughout this book, the role of situational factors must be taken into account when one judges the meaning of any nonverbal sign. For example, we may not be able to generalize from the results of studies in which observers rated the filmed behavior of therapists to the actual experience of those in a therapy session. Fehr and Exline (in press) noted several studies which have revealed discrepancies between observers' ratings of nonverbal behavior and the ratings of those actually involved in the interactions (Ellsworth & Ross, 1975; Holstein, Goldstein, & Bem, 1971). One study (Fretz, Corn, Luemmler, & Bellet, 1979) found that observer ratings of a counselor's behavior did not "hold up" in the actual counseling session (p. 123). Brill (1983) pointed out that most studies about the effectiveness of counselor behavior do not reflect the actual client populations to which we would like to generalize the results.

When the therapist tries to convey a message with the eyes or interpret the meaning of another's eye behavior, s/he must keep in mind such varying client characteristics as cultural and subgroup background, individual differences, relative status, etc., as well as contextual variables such as topic of discussion, physical proximity, seating position, setting, etc. Emphasizing the role of the client's unique subjective point of view in determining the effectiveness of counselor behaviors, Brill (1983) suggested that counselor education include training in the identification of clients' negative reactions to various counselor behaviors. She proposed the viewing of videotaped counseling sessions to aid the counselors-in-training to adjust their behaviors so as to develop their positive impact on clients. Brill's (1983) study also indicated that FIRO-B ratings on interpersonal orientation to affection, control, and inclusion can be used as effective indicators of individual eye contact patterns. For example, clients preferred a high level of eye contact if they were high scorers on FIRO-B interpersonal orientation. Pamela Brill's (1983) study is an excellent model for future researchers; hopefully it reflects a trend toward awareness of the role of situational variability and client differences in the therapeutic use of eye contact.

# GROUP THERAPY

Eye contact can play an important role in the process of group psychotherapy. Winick and Holt (1962), in their observations on the use of gaze in psychotherapy groups, reported that subgroups within the total group could be established by means of eye contact. For example, joint hostility toward another member or affinities for certain members of the group could be shown. The researchers found that looking at objects was a tactic to avoid looking at other people, and that looking upward might be an indication of inner awareness which occurred while others were talking about the person in question. It was also noted that winking and looking at sex organs were used as seductive signals. Callan, Chance, and Pitcairn (1973) also reported on the use of gaze as a social signal in therapy groups. Gaze apparently plays a central part in four types of display: (1) "Alert" people gaze at others and are open to interaction. (2) "Huddled" people avoid visual contact and withdraw from the group. (3) "Closed" people keep their gaze averted. (4) "Away" people have an unfocused gaze into space.

In a psychotherapy group, eye contact or its lack not only indicates something general about a person's emotional state but also indicates the focus of a client's interest. Often clients tend to look at the therapist more than at other clients, as a way of recognizing the status or authority which is vested in the professional. When one group member looks at another, the look may reflect an expression of closeness or empathy.

The therapist may often try to foster eye contact among group members as a way of promoting respect and a feeling of self-worth and relatedness among group members. As the client begins to make more eye contact with group members, the other members will probably respond more to the client. Since there is a direct correlation between liking and eye contact, the more eye contact a person engages in, the more she will be favored in the group. When a client looks mainly at the therapist, there is an unspoken message that it is only the therapist to whom the client wishes to communicate. The group members may then lose interest in what the client is saying. The therapist may become uncomfortable as well, with the obvious overinvestment of energy in herself. People sometimes have to be told that they are looking only at the therapist and then they may consciously attempt to focus more attention on others. Another way of discouraging a client's exclusive eye contact with the therapist is for the therapist to refrain from returning the gaze. Previously discussed research has shown that eye contact on the part of one person encourages eye contact on the part of the other. Thus, a therapist may naturally engage in a lot of eye contact in order to encourage this contact from others.

One way that a therapist can encourage sensitivity and awareness in contact among group members is to lead eye contact exercises in the safe environment of the group (Blank, Gottsegen, & Gottsegen, 1971). In a movement exercise, participants walk slowly toward one another across a long room, maintaining eye contact with a partner. Group members in the exercise report a feeling of "connection between our bodies and movement" or a "commitment between us." Bioenergetics exercises often involve eye contact. In one exercise which is used to train people to become more assertive and expressive of anger, people are instructed to stand up, make eye contact, stamp on the floor, and at the same time point to the other person with one hand while shaking the other (as a fist) saying in a loud voice, "I'm right. You're wrong." Some people are so threatened by anger that they cannot maintain eye contact while participating in this exercise, even if they are able to manifest anger well in other ways. When it is pointed out that they must make eye contact, there may be a great change in their reaction to the exercise.

A common exercise sometimes used at the beginning of a group may include the leader's request that participants gaze intently into one another's eyes, perhaps that they communicate specific emotions, (e.g., anger, love, caring, concern, fear, surprise, indifference). The exercise can vary from one of self-expression ("express your deepest feelings") to one of empathy ("experience the feelings of the other person"). If a group member makes eye contact and at the same time expresses a deep feeling to each other group member, emotional intensity and a sense of safety and trust are greatly enhanced (Ballantine, 1984). To give an example, a group member goes around the room, looks into each person's eyes, and says "I'm scared." Initially, people experience discomfort and self-consciousness in these eye contact exercises, which may cause them to giggle, talk, or avert gaze. These same exercises may take place while looking into a mirror instead of into another's eyes. There is a great difference in expression if one stares in an indifferent, empty way or with an interested, caring look (Hammond, Hepworth, & Smith, 1977).

Eye contact also has been used for an exercise in which group members stare into another's eyes in silence. This prolonged direct contact may provoke the release of intense emotions. An exercise described by Jack Lee Rosenberg (1973) involves two people sitting with joined hands, facing each other, and engaging in silent "eye-alogue." They look into each other's eyes trying to be "present" to each other with their "whole beings." Another technique has two persons sitting face-to-face about two feet apart. The first person intently observes the second person and agrees that s/he will respond to the second only when the second is communicating something through eye contact, such as "touch me," "get out of my sight," "massage my forehead," "comfort me."

In one group session, members went through a series of nonverbal messages and discovered how their relationship and communication problems had been developing. One person was trying to send the eye message "respect me." The other person was confused and waited a long time without understanding the message. The first person then intensified the contact, which felt like anger to the recipient, who responded with hostility. The result was that one person was trying to say "respect me" and receiving hostility from the other person. When the two of them talked afterward, they discovered that at many times during the group session the second person had felt suspicious of the first person's demands and dislike of her. In reality, the first person was feeling unsure and asking for respect but not expecting to get it. Eye contact between these members of a therapy group facilitated awareness that a pattern of miscommunication existed and aided insight into the source of the conflicts which resulted from the miscommunication. By paying attention to the messages of the eyes, we can gain access quickly, directly, and intensively to knowledge about basic emotional issues, thus creating an opportunity for accelerated insight and consequent psychological growth.

## CONCLUSION

As this chapter shows, eye contact can play a significant role in the therapeutic process, whether it occurs between members of a psychotherapy group or between client and therapist. Mutual gaze aids in establishing a warm, supportive relationship and facilitates the communication of other therapeutic attitudes which are conducive to positive change. The skilled therapist also utilizes her or his observation of the client's use and amount of eye contact. For instance, the client's eye contact can be discussed as a cogent example of her/his style of relating to others, or be observed as a concrete gauge of the client's therapeutic progress. Clinicians value the observation of patterns of eye contact for the assessment of client problems. As a means of increasing the client's capacity to relate with other people, eye contact can be introduced through many kinds of exercises and techniques into both individual sessions and group psychotherapy.

Considering the wealth of information now available about the role of eyes in human encounter, it is hard to believe that the study of eye contact by psychologists had scarcely begun twenty years ago. Today, we accept eye contact (along with other aspects of nonverbal communication) as a valid variable in all kinds of psychological research; clinicians use eye contact in diagnosis and therapy; and the human potential movement and other expressions of humanistic psychology have incorporated eye contact as a favored technique for the facilitation of interpersonal communication.

## FOOTNOTES

[1]The author refers to people who employ the services of psychotherapists as "clients" rather than as "patients" in the hope that the word will emphasize the equal status of such persons and discourage viewing them as "sick." To consider people's troubles in living as "mental illness" fosters the belief that the "disease" can be "cured" just as soon as the correct medical prescription can be found. This concept of illness obscures our understanding of how "conflicting personal needs, opinions, social aspirations, values, and so forth" (Szasz, 1973, p. 8) contribute to life's problems. The idea of mental illness also causes stigmatization, with the illness being seen as a "deformity of the personality" (Szasz, 1973) and the person who "has" it is seen as abnormal, a freak.

[2]At present, the validity of some of Bandler and Grinder's belief lacks conclusive experimental support.

[3]The person who needs and likes people tends to look more at others, but not in all situations. Exline (1963) found that competition inhibited eye contact among high-need affiliators. Since mutual gaze is a reciprocal and intimate act, highly affiliative persons may tend to reduce their eye contact because of the constraints of a competitive relationship (which is neither reciprocal nor intimate). They might also have a greater need to avoid cues of rejection and antagonism. Research on the relationship between individual differences in affiliative needs and patterns of eye contact is not altogether conclusive. Some studies (e.g., Exline, Gray, & Schuette, 1965) provide evidence to support the findings cited above. Others (e.g., Kendon & Cook, 1969; Gray, 1971) have found no relationship between affiliation orientation and looking behavior.

[4]It should be noted that the relationship between gaze avoidance and high levels of arousal has not been definitely proven, according to some researchers. Hobson, Strongman, Bull, and Craig (1973) concluded that gaze aversion was not a function of anxiety, whether transient states of induced anxiety or relatively permanent predispositional anxiety. There also has been much controversy over the Equilibrium Theory (see Chapter III), which postulates a relationship between intimacy, high arousal, and gaze aversion.

# CHAPTER VI

# *Eye Symbolism*

There is something about looking into another pair of eyes that is an intense experience. Although social science and clinical practice provide us with knowledge about how eye contact functions and how it may be used to enhance self-revelation, the psychological literature reveals little about the rich background of eye symbolism which has been a part of the heritage of countless generations all over the earth. Our understanding about why we react as we do to eyes and eye images must include knowledge of the ideas we have developed about eyes, ideas which are expressed by the cultures which surround us as our consciousness is formed.

We are an eye-conscious people. There are countless expressions which involve the eyes: "see eye to eye," "have an eye to," "give (a person) the eye," "an eyesore," "an eye-opener," and so on. Popular song titles, old and new, reveal our preoccupation with the eyes: "Ma, He's Making Eyes at Me," "I Only Have Eyes for You," "Sexy Eyes," "Girl with the Hungry Eyes," "My Eyes Adored You," "Spanish Eyes," "Bette Davis Eyes."[1]

Numerous words are used in ordinary speech to distinguish the various postures and often subtle movements of the eyes. For example, variants of the verb "to look" include: to stare, watch, peer, glance, peep, glare, gaze, contemplate, and scan (Gibson & Pick, 1963). A comparison of the occurrence of the words "mouth" and "eye" in *Bartlett's Familiar Quotations* (1968) shows that "eye" recurs four times as often as does "mouth." Similarly, a word count taken of normal conversation shows that "eye" is used more frequently than any other part of the body (Howes, 1966). In many theatrical dramas, the first look that occurs between two characters is often crucial to the unfolding plot (as in the song, "Some Enchanted Evening"). There is a whole book in French which is written on the topic of the first look in a play (*The Divine Eye*, Rousett, 1981).

What gives the eyes and eye contact power in our lives? Mythology and religion reveal that the symbol of the eye is a primal one spanning centuries and continents. Shamans, medicine men, and voodoo doctors through the ages have relied on their eye contact to practice their healing methods. Superstitions, folklore, and custom often involve the eyes. Writers, painters, sculptors, and other artists have used eye symbolism to convey meaning and express their understanding of the world. This chapter will explore the meanings of eye symbolism throughout history and across cultures by examining the role of the eyes in mythology, religion, the arts, language, literature, folklore, and custom. As we discover the sociocultural heritage each person brings to her/his encounter with an eye image or with another pair of eyes, we may also clarify the meanings eyes and eye contact can have in our lives.

## THE DIVINE EYE

There are many parallels among the various mythologies of the world, despite geographical and cultural differences. Certain symbols and the ideas they evoke surface throughout the world and throughout time, forming a kind of global language of shared archetypes.

Bachofen (1967), Graves (1948), and Neumann (1974) have documented that one of the earliest of these archetypes, the first deity in human history, was a goddess, the "Great Mother." Images of her have been unearthed by archaeologists around the world, and among these artifacts we find the eyes used as a symbol of divinity.[2] As Dames (1976) pointed out: "The eye motif played an orthodox and vital part in "Great Goddess" worship wherever Neolithic culture developed; certainly it was important in Spain, France, and the British Isles. In these places, the oculi designs were integrated so thoroughly as to become wholly characteristic, appearing on pots, portable idols . . . and engraved on the walls of tombs" (p. 67).[3]

In his study of Europe's tallest prehistoric structure, Silbury Hill in Great Britain, Michael Dames (1976) described one of the most fascinating and monumental examples of goddess/eye imagery. Seen from the air or from the East, the shape of Silbury Hill and its surrounding lake-quarry is the outline of the squatting mother goddess. Seen from the West, however, the goddess is all eye, head, and neck. Silbury Hill is "the biggest eyeball in Britain, set in the watery head of the Great Goddess" (p. 57). The womb of the Goddess from one direction becomes the eye of the Goddess when seen from the opposite direction. The hill is made of quarried chalk, its color thus reflecting the color of the human eyeball, with a round indentation at the summit to emphasize the pupil.

Dames (1976) suggested that the Silbury womb/eye synthesis represents the matriarchal Neolithic idea that mind and body, spirit and matter are unities. There are examples of eye/womb fusion from other cultures as well: e.g., the generative eye pictured in the Sumerian figurine from the third millennium B.C.[4]

Art historian Helen Gardner (1980) pointed out how "one notices at once . . . how disproportionately large the eyes are" (p. 43) in a sculpture of a female head from Warka, Sumeria. Gardner (1980) speculated on the meaning eyes might have had for ancient artists: "Large eyes fixed in unflagging gaze see all, and frontal, binocular vision, distinguishing human from mere animal seeing, represents the all-seeing vigilance and omniscience of the gods and the guarantee of justice" (p. 43).[5]

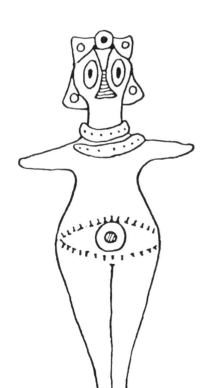

FIGURE 6-1 The "generative eye," Sumerian figurine from the third millennium B.C. (Reprinted with permission from *The Silbury Treasure*, by Michael Dames. London: Thames and Hudson, Ltd., 1976.)

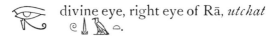

divine eye, right eye of Rā, *utchat*

divine eye, left eye of Rā.

the two divine eyes, *utchatti,*
, the eyes of Rā, *i.e.,*
the Sun and Moon.

FIGURE 6-2 The "divine eye" appeared in Egyptian hieroglyphics. (From *An Egyptian Hieroglyphic Dictionary* by Sir E. A. Wallis Budge. New York: Dover Publications Inc., 1978. Reprinted by courtesy of the Master and Fellows of University College, Oxford, and the Christ's College, Cambridge.)

In Egypt, the cat goddess Bastet reigned for about 2000 years. The origins of her power seemed to rise from the magical properties attributed to cats' eyes (Cantin, 1976). The eye was a very common symbol in ancient Egyptian thought, pervading the monuments and sacred texts of all periods (Talbot, 1980). The hieroglyphic "Utchat" was one of the most common names for the eye (Talbot, 1980). Although the sun gods Osiris and Ra were often associated with an eye, a close examination of Egyptian texts reveals that the gods are not the eye itself. Rather, the god is born from the eye (Budge, 1969) or dwells within it (Talbot, 1980). The eye is, again, the womb of the goddess, "a female personification of the watery matter which formed the substance of the world . . . a form of the primeval female creative principal . . ." (Budge, 1969, Vol. I, pp. 422–423).[6]

In many cultures, the sun was associated with the divine eye, a luminous and powerful force watching over humanity with an untiring gaze. In sacred Hindu writings, the sun is called "the eye of the world," or Toka-Chakshun. Helios, the Greek god of the sun, was referred to as the "eye of the day." Christianity preserved this association by calling Jesus the "light of the world," and for many centuries Christian art represented the all-seeing, all-powerful God as an eye (or a hand or an arm) emerging from the clouds (Hulme, 1969).

The advent of Christianity did not shake the divine eye's hold on the imaginations of many peoples. In Mexico, the eyes of the rain god Tlaloc have persevered for centuries as onyx inlaid in pyramids built for him. Catholicism incorporated the divine eye by placing it within a triangle representing the Holy Trinity. A triangle with an eye in it can be found on the domes of churches and on walls in courtrooms throughout Mexico today. Silver eyes on small plaques are sold in the marketplace to be given as offerings at the churches or to be worn as necklaces, earrings, or rings. In an article tracing the eye motif in art, Katherine Kuh (1971) noted that: "Even today Mexican Indians from Nayarit celebrate a certain festival with objects called 'God's eyes,' which in reality are woven reed wands decorated geometrically with brightly colored abstract eyes. According to legend, these wands

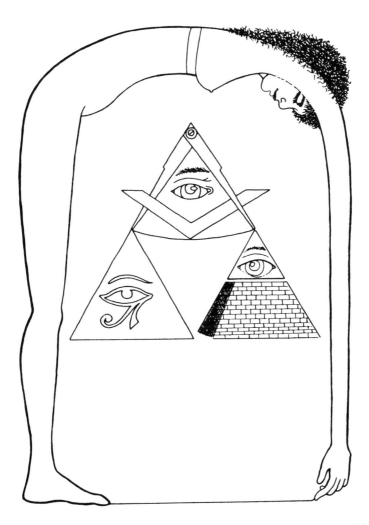

FIGURE 6-3 Three forms of the eye-in-the-triangle design. The one on the left is the emblem of Aleister Crowley's magik society, the Ordo Templi Orientis. The one on the right is the Great Seal of the United States. The top one appears in every Masonic Lodge. (Reprinted with permission of And/Or Press, Inc., from the book *Cosmic Trigger* by Robert Anton Wilson. New York: Pocket Books, 1977.)

encourage one of their local gods to see more clearly and to understand more fully the mysteries of life" (p. 153).[7]

Today, the divine eye can be seen in Greek Orthodox churches in the United States; on the ceilings of houses in Bavarian and Austrian villages; on the back of the one-dollar bill; and on Buddhist monuments in Nepal. In homes and public places around the globe, "god's eyes" are hung as a protective measure. It is a universal symbol that has remained a powerful reminder of humankind's idea that there is an ever-watchful deity.

Secular art has incorporated the abstract symbol of the all-seeing eye as well. The eye is a magic symbol in the surrealistic paintings of Ernst, Dalí, and Magritte (Martí-Ibañez, 1952). In Magritte's surreal closeup oil on canvas, "The False Mirror," an eyeball floats in an orbit composed of sky and clouds: "The picture shows inner and outer worlds inseparately joined. One has the feeling of looking inward, of looking inside another person; at the same time one has the feeling of looking outward at the world. Suddenly one becomes the person in the picture, staring into and out of his own eye" (Samuels & Samuels, 1975, p. 76). This journey into many levels of seeing through the image of an eye seems linked to the universal use of eyes as symbols of divine omniscience. However, this painting invites us to

**FIGURE 6-4** *The False Mirror* by Rene Magritte (1928). Oil on canvas, 21¼" × 31⅞". Collection, The Museum of Modern Art, New York.

an awareness of our own powers of vision and could thus be interpreted to include humanity in a concept of the divine.

Literature also provides abundant evidence that eyes have divine significance, even for the modern mind. In *The Great Gatsby* (1925), F. Scott Fitzgerald finds such eyes in an unexpected place, a billboard advertisement in a "certain desolate area of land" near New York City:

> ... above the gray land ... which drift endlessly over it, you perceive, after a moment, the eyes of Doctor T. J. Eckleburg. The eyes of Doctor T. J. Eckleburg are blue and gigantic—their retinas are one yard high. They look out of no face, but, instead, from a pair of enormous yellow spectacles which pass over a non-existent nose. Evidently some wild wag of an oculist set them there to fatten his practice in the borough of Queens, and then sank down himself into eternal blindness, or forgot them and moved away. But his eyes, dimmed a little by many paintless days, under sun and rain, brood on over the solemn dumping ground. (pp. 27–28)

The eyes of Doctor T. J. Eckleburg take on greater meaning when Wilson informs Michaelis of the suspicions he had of his wife before she died:

> "I spoke to her," he muttered, after a long silence. "I took her to the window . . . and I said 'God knows what you've been doing. You may fool me, but you can't fool God!'"
>
> Standing behind him, Michaelis saw with a shock that he was looking at the eyes of Doctor T. J. Eckleburg, which had just emerged, pale and enormous, from the dissolving night.
>
> "God sees everything," repeated Wilson. (pp. 191–192)

Half-consciously, Wilson identifies the eyes of Doctor T. J. Eckleburg with God. This "vulgar transformation of the image of the all-seeing Deity" (Fowler, 1966, p. 23) makes the reader aware of "some kind of detached intellect brooding gloomily over life . . . and presiding fatalistically over the little tragedy enacted as if in sacrifice before it" (Miller, 1968, p. 36).[8]

A corollary to the use of the eyes as representations of divinity is the idea that supernatural powers are reflected or located in the eyes. In Greek mythology, Zeus, all-knowing and far-seeing king of the gods, could cause lightning with his glance. Athena, his daughter and goddess of wisdom, was described as "flashing-eyed" (Hamilton, 1969, p. 29). Aphrodite, goddess of love, had "sparkling eyes" according to Hesiod (trans. by Brown, 1953, p. 53). Large eyes were painted on the outside of Greek wine cups, perhaps to help ward off the evil effects of drinking too much wine. With a glance of his eye, the Indian god Siva was said to have burned to ashes Kama, deity of love (Keith, 1917).[9]

**FIGURE 6-5** Cedar bark hat with eye designs. Tlingit, Alaska. (Photograph courtesy of Museum of the American Indian, New York, Heye Foundation.)

Variation in the number of eyes has been used to represent special powers. Supreme watchfulness was epitomized in Argus, the hundred-eyed creature of Greek legend. Only two of its eyes ever slept at one time. The Indian sky god, Varuna, whose eye was the sun (Mitra), was said to possess one thousand eyes and was thus undeceivable (Keith, 1917). The Cyclops were a mythical race of giants with an eye in the middle of their foreheads. Their single eye symbolized their awesome power. When blinded, the Cyclops became helpless, despite their supernormal strength. The Gorgons were three wise women who shared a single eye between them and whose glance could turn the beholder into stone.

The divinity of eye symbolism is further related to the concept of the eyes as containing the power of life and death. The "Queen of Heaven" tombstone, with its cosmic symbols for eyes, showed the eyes' ancient association with the deity who ruled over the world of the dead. Indeed, in our own experience, life seems to begin as the eyes open; the finality of death is marked by their closing. A child may believe that if her eyes are kept open s/he will not die. The solipsist might argue that closed eyes become an instrument of death.

As mythology, archaeology, art, and literature show, eye images have been used by many cultures throughout history as symbols of divine presence and power. Originally associated with goddesses, the symbol of the divine eye survived major shifts in social organization—from matriarchy to patriarchy, as well as from the belief in many gods to the belief in one. Because of its universal significance, artists and writers have used the image of the divine eye to evoke the presence of an observing god as well as to convey magic, mystery, and power.

An essential element of the constellation of meanings associated with the divine eye is the eye as a symbol of special power. As described in this section, such power can inspire reverence and awe. But what if such power is used destructively? Then the eyes become fearful signs of evil, over which there is no control. Fear of that kind of power has surfaced in the consciousness of humanity as the Evil Eye, the "flip-side" of the divine eye and just as enduring.

## THE EVIL EYE

In the film version of "The Wiz," the facade of the Wizard of Oz exuded smoke from the nose and light streamed from the eyes. The streaming light made the eyes appear to have great powers and to transmit an almost tangible force. Picasso's gaze was described as being so intense that "whenever he looked at a picture one was surprised to find anything left on the paper afterwards" (Martin, 1980, p. 18). Theosophical scholar G. de Purucker (1974) asserted that actual visual rays do emanate from the eyes. He said that these rays, or electromagnetic forces, are especially powerful when "the will is behind and propelling the personal auric magnetism" (pp. 6645–6646). According to de Purucker, looking a person directly in the eye is indeed admirable, for to do so involves a brave "battle of magnetisms."

Our consideration of the "divine eye" makes it clear that in both popular culture and mystical traditions the eyes are often perceived as sources of power. It follows that the eyes may also be feared as sources of unwanted influences over ourselves. We may dread their control. As Nancy Henley noted in *Body Politics* (1977): "We are half-frightened by the prospect of being controlled by someone's eyes, in the popular image of the 'Svengali' who intones, 'Look into my eyes

... Look deeeeep into my eyes' ... a traditional acting-out of the archetypical scene of external control, the hypnotist or mad doctor exerting power over the innocent subject" (p. 152).

Phyllis Greenacre (1926) noted that some emotionally-disturbed clients have delusions that power is exerted upon them and enters their bodies through their eyes or that their own eyes are able to control the activities of others. The idea that one emanates influence from the eyes is often associated with a sense of guilt, and the emanating influence is felt to be an evil or malicious force. One woman felt her eyes could cause others to go blind or insane (Greenacre, 1926, p. 562).

Images of the eyes can be haunting, causing us uneasiness and discomfort:

> From the hour we are born until the day we die they are upon us, watching us always.
> The eyes. Loving, kindly, pitying, agonized eyes. Cold, jealous, scornful, evil eyes.
> Watching us always . . . .
>
> They follow us in the streets, cold eyes,
>   condemning eyes, noticing our little awkwardnesses, roaming over our garments, waiting a movement to ridicule—hard, uncharitable eyes. From the dawn until the dark they pursue us with their attention.
>
> Let me get away from the eyes . . . (Sigerson, 1920, p. 57)[10]

Ocular images have a way of creating "an indefinable sense of guilt," as Kuh (1971, p. 152) observed, especially if they are disembodied and placed in a repetitive pattern. As Kuh noted, "The eyes become a human compendium, meting out final judgments so impartial as to disquiet even the blameless" (p. 152). For those who have reason to feel guilty, staring eyes can be horrifying.[11] Edith Wharton punished the egotism of Culwin, a bachelor in her short story "The Eyes" (1910), by the nighttime hauntings of a pair of eyes. Awakened by a "queer feeling," Culwin said,

> I strained my eyes into the darkness . . . as I looked the eyes grew more and more distinct: they gave out a light of their own . . . They were the very worst eyes I've ever seen: a man's eyes—but what a man! . . . The orbits were sunk, and the thick red-lined lids hung over the eyeballs like blinds of which the cords are broken . . . What turned me sick was their expression of vicious security . . . as we continued to scan each other I saw in them a tinge of derision, and felt myself to be its object. (pp. 253–255, passim)

Culwin was unable to understand the origin of these eyes until he was finally forced to recognize that the eyes he was so terrorized by were his own.

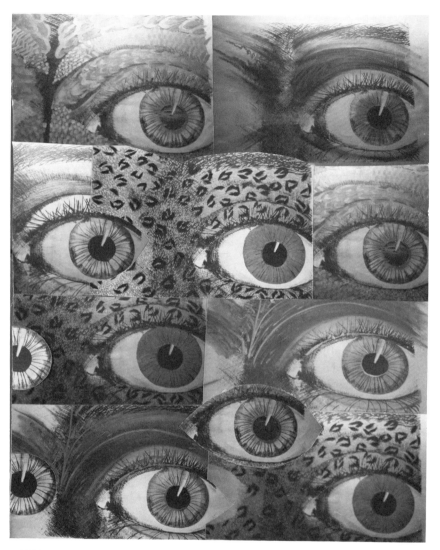

**FIGURE 6-6** *The Observers.* Collage designed for this book by Michael Gaines.

Although Wharton's story is fiction, it indicates the frightening meanings which can be ascribed to eye images. Sometimes these fears will surface in our dreams. Andrew Peto (1969) noted that the appearance of terrifying eyes in one's dreams can represent archaic superego elements of the personality. The angry red glow of a parent's disapproving, frightening eyes is the cause of shame and guilt in early childhood and becomes internalized as symbolic representations of the superego. This kind of association helps to clarify why images of the eyes have their ability to haunt us—they can be symbols of our own conscience, our internal source of guilt and self-punishment.[12]

With the increasing use of citizen surveillance by governments, a practice not unknown in even presumably democratic societies, the eye has become a symbol of intrusion, a violator of privacy and individual autonomy. The symbol of the CBS television network, a large eye, has often been used by political cartoonists to represent government surveillance. In George Orwell's world of 1984, citizens are watched continuously by the all-seeing eye of "Big Brother," through two-way television sets in every home, and from the ubiquitous posters whose eyes always stare directly at the viewer.

In many cultures it is believed that a glance can cause sudden harm to another's property or person (Tomkins, 1963). This is the commonly known power of the "evil eye." Those thought to possess it are imagined to shed rays of destruction about them with every glance, frequently unaware of their dread influence (Trachtenberg, 1974).[13] Many explanations from folklore "maintain that the hurtful effect of the eye results from feelings of jealousy, envy, greed and frustration on the part of those obliged to admire the good fortune . . . of another" (Teitelbaum, 1976, pp. 64–65).[14]

The Catholic Church institutionalized the evil eye as a sign of witchcraft when Pope Innocent VIII validated the *Malleus Maleficarum* with a Papal Bull in 1484. Written by Fathers Heinrich Kramer and James Sprenger, the *Malleus Maleficarum* (1971) stated in detail what the signs of witchcraft were and how witches were to be prosecuted, tortured, and executed. They warned that the devil could creep into human bodies through the senses and, once inside, could turn the most direct and concentrated of those senses, the eyes, into his weapon against unsuspecting, vulnerable souls, particularly children. The evil eye and the inability to shed tears were among the signs that one was a witch.

Kramer and Sprenger's writings spread throughout Europe, fueling the witch craze of the 15th, 16th, and 17th centuries. During a period of 300 years, 9 million people were put to death as witches (Gage, 1980). Many scholars have determined that belief in the evil eye and witchcraft are closely related, and some have argued "that the evil eye is only a special case of witchcraft belief" (Garrison & Arensberg, 1976, p. 319). Like accused witches, those suspected of casting the evil eye are most often women. Menstruating women, for example, have been abhorred and thought to be "like the viper that kills with its glances" (Trachtenberg, 1974, p. 185). As one might expect, belief in the evil eye is found basically in patrilineal cultures (Spooner, 1976), and has likewise served to keep social stratification justified and intact through blaming tragedies on the less tolerated people within a society: women, poor people, beggars, the disabled, outsiders, and people who deviate from accepted behavior. It was widely believed by Christians in Europe that Jews could cast an evil eye; a German word for evil eye is still *Judenblick* or Jew's glance (Moss & Cappannari, 1976).

Fear of the envy of others seems to have been a major contributing factor to the continued existence of belief in the evil eye (Maloney, 1976). In cultures where much is to be gained by greed, and yet greed is considered evil, belief in the evil eye flourishes. According to anthropologist John M. Roberts (1976), "the evil eye becomes prominent in a culture when the society produces goods that can be envied and when . . . there is an unequal distribution of these goods in the presence of social inequality" (p. 238).

As we have seen, eyes and eye images emerge out of the vast symbology of human existence not only as symbols of divine power, but also as haunting images which can terrify us in our dreams and which are used to scare us in science fiction and horror films. Perhaps our frightened responses are related to an innate aversive response to staring eyes inherited from the animal world, in which the mutual glance is a form of threat (Argyle & Cook, 1976). Koenig (1975) observed that the eye motif is prevalent in human crisis situations. He concluded that the frequent use of the eye symbol by a society signifies a disturbed, troubled societal structure.

The sometimes haunting nature of eyes and their images, combined with fear of them as sources of uncontrolled influence and destructive power, has contributed to belief in the evil eye around the world. The eyes seem to lend themselves easily to projection. Somehow, people are apt to see their own guilt, envy, greed, and other hostile feelings reflected in the eyes of others (Hart, 1949). Peto's (1969) psychoanalytic interpretation of eyes in dreams emphasized that eyes can be symbols of our superego—a part of the self watching over the self. Indeed, eye images are associated with aspects of the self, with one's very identity.

## I AM EYE

Language is another aspect of culture which reveals the connections between certain ideas and eye symbolism. Native Americans sometimes employ the word "eyes" in the names they give their children,

such as "Bright Eyes" or "Iron Eyes." In the unconscious language of
dreams, it has been found that "eye" and "I" are often equated
(Huebsch, 1931). In English the words "I" and "eye" are pronounced
identically, and in Greek they are spelled identically as well: "ai." In
the romance languages, they have similar sounds: *io* and *occhio* in
Italian; *yo* and *ojo* in Spanish; and *je* and *yeux* in French.

For "eyes" to be used as names, for an eye to represent oneself in a
dream, for "I" and "eye" to sound the same or similar in several
languages, show that eyes and eye images are linked with the concept
of identity.

Popular belief through the ages has told us that the eyes reflect a
person's true nature and emotional state—that they are openings
through which others can see into us. Da Vinci referred to the eyes as
"the window of the soul" (Hart, 1949), and others have variously
called them "windows to my breast" (Shakespeare, 1919), "beloved
little windows" (Keller, 1960), and gates to the "depths of our being"
(Rosetti, 1970). Thomas Adams, 17th century Puritan clergyman and
writer, expressed the prevailing sentiment about the eyes when he
noted, "The eye is the pulse of the soul; as physicians judge the heart
by the pulse, so we by the eye" (Bradley, Daniels, & Jones, 1969,
p. 272). Willner (1968) asserted that the eyes, more than other parts of
the body, are indicators of personal qualities or character. "Nobody
would get the idea that he could look into another man's 'soul'
through the ears, nose or mouth . . ." (Katz, 1963). Recognized as a
key to individuality, the eyes are masked by criminals wishing to
disguise themselves; they are blacked out in scandalous photographs
to protect the subject's identity; and sometimes an eye-mask alone is
considered disguise enough for a costume party.

What experiences with the eyes contribute to the idea of identity
and illuminate the source of the connection between "I" and "eye"?[15]
For sighted persons, eyes are a major tool for discovering and exploring the world. Eyes, therefore, can be equated with one's identity as an
observer. Also, realizing that one is not that which one observes
contributes to a sense of a distinct self. All persons, sighted or not, are
perceived through the eyes of others. One's identity is thus taken in
by others' eyes and we can see ourselves reflected in them (literally
and figuratively). The process of "seeing" and "being seen" reveal our
identity to ourselves and to others.[16]

Visual metaphor can also be a powerful means of expressing our
own self-consciousness, the struggle to perceive ourselves as we really
are. In the novella *The Eye* (1965), Vladimir Nabokov explored the
haunting desire of an unhappy, self-obsessed young Russian named
Smurov to encounter his true identity. Nabokov used eye imagery to

represent Smurov's obsessive self-consciousness which had become a constant torture:

> I was always exposed, always wide-eyed; even in sleep I did not cease to watch over myself, growing crazy at the thought of not being able to stop being aware of myself. . . . (p. 17)

Eventually Smurov realized that he could not depend on the ideas others had about him, on the images other eyes held of him. But this knowledge did not throw him into despair. On the contrary, it thrust him into an appreciation of his own vision and a claiming of his identity as an observer.

> I swear I am happy. I realized that the only happiness in this world is to observe, to spy, to watch, to scrutinize oneself and others, to be nothing but a big, slightly vitreous, somewhat bloodshot, unblinking eye. . . . I am happy that I can gaze at myself, for any man is absorbing. . . . (p. 114)

To be a writer requires a special kind of detached awareness which Smurov embraces with relish in the end. Thus, if we look at this story as "an early self-portrait of the artist who wrote it" (Unger, 1974, Vol. III, p. 251), the image of the ever-observing "unblinking eye" becomes a symbol of Nabokov's identity in the world: the observing writer.

As the languages of several cultures and the language of our dreams reveal, the eyes are linked with the concept of identity. Seeing is part of the development of our individual perspectives on the world; being seen is crucial to the revelation of our selfhood to others and subsequently to ourselves.

But "to see" does more than simply provide us with a sense of self; it is vital for the knowledge it provides us about the world we inhabit. For this reason, the eyes are used as symbols of knowing, and a wealth of mythology and literature has relied on this metaphor, as we shall see in the next section.

## I SEE, THEREFORE I KNOW

There are many ways in which the eye is used as a symbol of mental perception. At the end of a weekend-long therapy group session, a woman sculpted an eye out of clay, explaining that it symbolized the insight and awareness she had gained through the experience. Sets of eyes are used on the covers of psychological journals and books to imply a message about the mind—about the way information is taken in and understood by the "mind's eye." On the covers of educational books, eyes are often portrayed to suggest the quest for knowledge.

It is often said that "seeing is believing." "Do you see?" is a common way to ask, "Do you understand?" When we refer to someone as "having her eyes open," we mean that s/he is well aware of the situation and will not overlook valuable information. Such everyday usage of eye images and visual metaphor reveal that the eye is commonly used as a symbol of knowing. Indeed, in English, the word "idea" (thought, mental image, notion) has its roots in the Greek work *idein*, which means "to see" (*Webster's*, 1968). The books of the ancient prophets of India are called the *Vedas*, a Sanskrit term that relates both to vision and wisdom (Arguelles & Arguelles, 1972).

Sighted infants learn by continually looking upon new things they had not known existed. By mentally assimilating new visual input, children grow in competency and understanding. Sometimes, however, our compulsion to look will give us knowledge we wish we did not have. How many times have you kept your eyes open in a horror movie when you knew a gruesome scene was coming, and then regretted it? Indeed, humanity's relationship with the unknown seems to be full of an intense desire-to-know on the one hand, and an equally intense fear-to-know on the other. Some sights bring joy and delight, others may bring confusion or horror. The anguish of perception, caused by cognizance of the more woeful aspects of the human condition, has surfaced as a theme in many myths about the "forbidden look." In such myths, those who dare to glimpse forbidden precincts of knowledge are punished.

Forbidden by God to partake of the fruit of the tree of knowledge, Eve dared to taste it, for the serpent told her, "your eyes shall be opened: and you shall be as Gods, knowing good and evil" (Genesis 3:5). The Greek myth of Pandora predates the Genesis story and is another well-known example of this theme. Pandora could not resist opening the box (or jar, in some versions) she had been forbidden by the gods to open. When she removed the lid to look inside, diseases of the body and spirit were unleashed on the human race, causing sorrow, misfortune, and death to enter the world. The Pandora myth uses the metaphor of the forbidden look in a more central way than the Eden story. Rather than tasting fruit (which then opens the eyes), it is the act of looking itself which is the source of evil in the world.

Looking is not only negative. In fact, using one's eyes can give much pleasure, even sexual pleasure at times, and sexual knowledge has long been relegated to the realm of the taboo in many cultures. Therefore, it is not surprising that the device of the "forbidden look" has been used to express the dynamics of ambivalence toward sexual knowledge. For example, Artemis, goddess of the hunt, turned a young hunter, Actaeon, into a deer after he stumbled upon her bathing in the nude. Athena did the same to Teiresias who chanced upon her taking a bath. (Later, however, she regretted her harshness and granted Teiresias the gift of inner sight or prophecy to recompense him for his loss.)

In another myth, the god of love, Cupid (son of Venus), fell in love with a beautiful, mortal woman named Psyche. To avoid his mother's wrath, he came to Psyche only at night and forbade her to look at him. Psyche's envious sisters persuaded her to light a candle and gaze upon her lover's face. When some hot wax from the candle dripped on him, Cupid awakened, decried Psyche's lack of faith in him, and fled (Hamilton, 1969). (It is interesting to note that Cupid was often represented as blindfolded, symbolizing the belief that "love is blind.")

As in the cases of Actaeon and Teiresias, blindness is used in some myths as the form of punishment for glimpsing the taboo. To use blinding in this way emphasizes the eyes as the symbolic cause of discovering forbidden knowledge. The Bible echoed this idea, "And if your eye causes you to sin, pluck it out and throw it from you . . ." (Matthew 18: 8, 9). When Oedipus heard from the oracle Teiresias the true nature of his fate—that he had unknowingly slain his father and married his mother—he tore out his eyes.

Greenacre (1926) suggested that a desire for self-blinding as an extreme form of retribution for sexual transgression may also represent a self-castration wish. De Vries (1974) suggested that Oedipus probably did not blind himself originally, but emasculated himself. Rank (1913) found through dream analyses that the eyes and genitals are often symbolically connected and they considered Oedipus' self-blinding to be a symbolic castration motivated by sexual guilt. Schlaegel & Hoyt (1957) hypothesized that since the sexual organs are taboo, emotion must be cathected onto another visible and highly valued organ of pleasure.

The self-blinding of Oedipus unites the idea of punishment for violating a sexual taboo with the idea of the anguish of perception—in this case, perception of his guilt for murdering his father and marrying his mother. Mad with horror from it, he extinguished the symbolic source of all knowledge, his eyes, saying:

> wicked, wicked eyes . . . You shall not see me nor my shame—not my present crime
> Go dark, for all the blind to what you never should have seen. (Sophocles, trans. by Roche, 1958, p. 73)

Male schizophrenic patients have been known to tear out both their testicles and their eyes in self-castration as punishment for fantasied forbidden pleasure (Schlaegel, 1957). A blind patient believed he was punished with loss of sight for visualizing sexual intercourse with his mother (Rank, 1913).

As the vehicles that bring light to darkness, the eyes are associated with wisdom and enlightenment. In the Bible, eyes are often used figuratively as the channels for spiritual knowledge: "And his eyes were enlightened" (1 Samuel 14:27); "That our God may lighten our eyes" (Ezra 9:8); "Lord is pure enlightening your eyes" (Psalm 19:8). The biblical scholar Milton equated light with eyesight and knowledge, dark with blindness and ignorance. It is also interesting to note the many cases of blindness that the Bible says Christ cured. Faith and knowledge of God seemed to allow the healing of blindness; i.e., in symbolic terms, spiritual ignorance.[17]

Despite all the negative associations made with defective vision and blindness, there is a thread which connects blindness and positive attributes as well. In Greek myth, for example, the true prophets and "seers" were blind. Remember Teiresias who was blinded by Athena and given the gift of prophecy in compensation. He appears in many stories, advising such characters as Odysseus and Oedipus. In Norse mythology, the supreme god, Odin, is in a constant search for wisdom. In one story, he went down to the "Well of Wisdom" and begged for a drink from its guard, Mimir the Wise. Mimir answered that he must pay for the drink with one of his eyes and Odin consented to make the sacrifice (Hamilton, 1969).

Max Ernst links blindness with clairvoyance in his picture, "The Blind Swimmer" (Samuels & Samuels, 1975). "Ernst has described himself as a 'blind swimmer,' who has made himself clairvoyant and has seen" (Samuels & Samuels, 1975, p. 252). His images emphasize the inner eye as a path of awareness. In his collotype, "The Fugitive," Ernst depicts a "fantastic eye traveling over the earth" (Samuels & Samuels, 1975, p. 278). According to Ernst, clairvoyance, symbolized by the roving eye, enables the artist to journey far from everyday reality.

If you take a minute to close your eyes and sit quietly, you can see that blindness can free one from distracting extraneous images of the outer world. It enhances the possibility of seeing with the mind's eye and of opening up knowledge from the inner self. With one's eyes closed, we can be more open to "seeing" with other than normal, physical sight, as in a dream or a trance; we can have "visions" (mental images) of things not actually visible. Perhaps this is why blindness, wisdom, and prophecy have been linked.

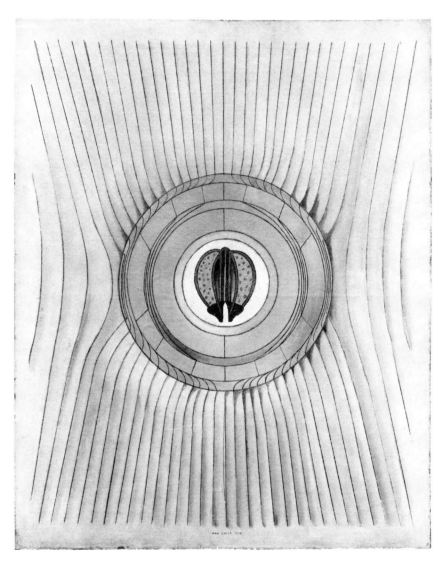

FIGURE 6-7 *The Blind Swimmer* by Max Ernst, 1934. Oil on canvas, 36⅜″ × 29″. Collection, The Museum of Modern Art, New York. Gift of Mrs. Pierre Matisse and the Helena Rubinstein Fund.

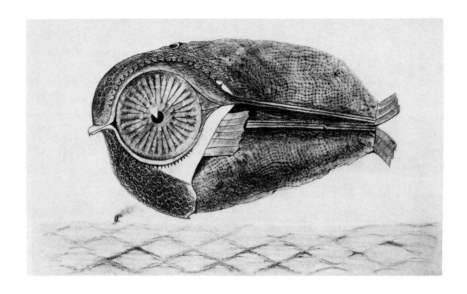

**FIGURE 6-8** *The Fugitive* by Max Ernst. From *Histoire Naturelle*, portfolio of 34 collotype reproductions after pencil frottages. 1926. Collotype printed in black, 16⅛″ × 10⅜″. Collection, The Museum of Modern Art, New York. Gift of James Thrall Soby.

In Asian art, lids are often partially or completely lowered. Kuh (1971) explores the symbology of the lowered lid noting, "The eye looks inward, for in the Orient salvation results from noninvolvement with the outside world. Men meditate, they shroud their eyes in silent secrecy . . ." (p. 156). As Kuh (1971) said, in European art the "open eye is central to our very being," showing Western emphasis on the material, outer world. However, artists can still represent an inward look with the open eye, as in Leonardo da Vinci's "Ginerva de Benci." Similarly, during the Hellenistic period of Greek art, in which sculpture begins to look more realistic with emphasis moving from the universal to the particular (Robertson, 1975), one starts seeing sculptured eyes which do not look at you, in which there is no contact. This lack of connection with the outer world evokes a sense of the inner and spiritual.

Like the lowered lid and the introspective open eye, the image of the "Third Eye" is a symbol of spiritual knowledge. Often pictured on the forehead above two normal eyes, the Third Eye represents a spiritual vision that transcends the logical, normal view and is the Buddhist symbol of inner wisdom. The Third Eye is considered a faculty of the astral, as opposed to the physical, plane (Gaskell, 1960). It is considered by many religious and philosophical traditions to be "the interior organ of vision," or the eye of Shiva (de Riencourt, 1981)

**FIGURE 6-9** *Subsequent Penetration into the Seventh Circle* by R. Lucassen. Collection, Stedelyk Museum, Amsterdam.

which can be directed to see through the illusions of our materiality to the essence of our existence—the reality of the god-within-us (de Purucker, 1974). The Third Eye is a symbol for that part of our mental perception which allows us to see ourselves and the world as an organic whole (Arguelles & Arguelles, 1972).[18]

The concept of the Third Eye implies a union of the powers of two eyes. One eye symbolizes immediate purpose without discrimination or intellectual perspective, and may also be seen as complete subjectivity, the "I" which represents no one else. With the addition of a second eye, one sees with perspective, detachment, and objectivity.

## CONCLUSION

Eye images and/or seeing have been used as metaphors of knowing in mythology, literature, art, psychology, and mysticism. Mythological characters were forbidden to look upon certain things to keep them from knowledge of the taboo; blindness often resulted from their daring to look. Blindness and poor eyesight have been (mistakenly) related to ignorance. However, blindness and inward looking have also been associated with "true seeing," wisdom, and the lively life of the "mind's eye."

As the foregoing review of mythology, religion, the arts, folklore, and custom illustrates, the eye is a constantly recurring symbol that is not confined to any particular culture or to any particular period of history. The eye has been used as a visible sign of concepts or abstract ideas such as divinity, power, clairvoyance, spiritual insight, emotional communication, etc. "Symbols are the alphabet of the mental plane. Energies and ideas are formulated into meaning through symbols, and then translated into 'understanding'" (Saraydarian, 1976, p. 182). Thus, eye symbolism has been used to keep alive certain understandings about human existence.

The association of eyes with the Great Mother, particularly the fusion of eye/womb imagery, evokes the mysterious origins of all life. The eye symbol is used in myth and art to represent spiritual understanding; it also is used by mystics to open the door to spiritual growth. The holy origins of the meaning of eye symbolism have certainly caused a connection between the eyes and the concept of power in the consciousness of humankind. And like many symbols of power, eyes as a visual metaphor have been exploited as tools to help maintain the prevailing moral order. For example, evil eye superstitions have assisted the scapegoating of socially "indigestible" groups in order to maintain a status quo, while the theme of the "forbidden look" in myth has taught sexual taboos.

FIGURE 6-10 Eyes are an important part of the design of these Shaman's charms. The top charm is made from a section of whale's tooth and was carved to represent three wolf figurines and a human on the right. Tlingit, Hootzahta Tribe, Admiralty Island, Alaska. The bottom charm is a bone carving representing a Spirit Canoe, Tlingit, Alaska, 1825–1875. (Photograph courtesy of Museum of the American Indian, New York, Heye Foundation.)

Although modern people may tend to laugh at or deny the belief that the eyes have power, hypnotists have observed that a "patient is more quickly hypnotized when he looks into the eyes of the hypnotist than he is when he merely fixes his gaze on some bright object" (Meares, 1960, p. 193). Since the eyes probably do not create an actual physical influence, the activation of the ancient belief in the influential power of the eyes provides a psychological explanation for this phenomenon. Meares also observed that "under stress of emotion or threat of loss of control . . . these ideas (about the eyes) seem to be capable of being reactivated and come to influence our thought, feeling, and behavior" (1960, p. 194). Meares believes that an unfriendly atmosphere can activate the undesired symbolic mechanisms of the evil eye. For this reason, he recommends that "ideas of the evil eye must be avoided at all costs . . . by establishing good rapport" (p. 195) in order to avoid fear and resistance to hypnotherapy.

Psychoanalysts also look to our symbolic heritage in their interpretation of dreams (Freud, 1955) and in their search for underlying causes of personality disturbances. As Hart (1949) asserted, "the eye has come through thousands of years of cultural evolution to have a symbolic meaning indispensable to the understanding of certain symptoms" (p. 1). Blank (1963) cited "the unconscious significance of blindness as castration, as punishment for sin" (p. 188) as one of the factors underlying the maladaptions and personality disturbances of visually handicapped persons. Greenacre (1926) noted many examples of how beliefs about the power of the eyes are expressed in the symptoms of those with emotional problems.

Since popular beliefs which link eyes and self-revelation still permeate our art and literature, one would expect that people will notice eye contact and search the eyes for knowledge of others. And, al-

FIGURE 6-11 Shaman's charm. Tlingit, Hoonah Tribe, Alaska, 1825–1875. (Photograph courtesy of Museum of the American Indian, New York, Heye Foundation.)

FIGURE 6-12 This ivory shaman's charm represents an animal head on one end, a raven on the other, and spirit figures between. Tlingit, Sitka, Alaska. (Photograph courtesy of the Museum of the American Indian, New York, Heye Foundation.)

though each culture provides a unique heritage, it is quite apparent that the power of the eyes fascinates people throughout the world. Hopefully, the knowledge reviewed in this book will enable us to embrace the language of the eyes as a valuable tool for communication and interconnection among human beings whose survival as a species depends on its ability to live in harmonious interdependence.

## NOTES

[1]Pop art also reflects a fascination with the eyes. There is a ready market for posters and shopping bags with giant eyes, eye mobiles, jewelry, and key rings with eyes (some of which even blink).

[2]In Neolithic times, the eyes of the goddess were sometimes portrayed in isolation and became "a formalized iconography of a goddess figure that assumed the role of the complete figure seen elsewhere" (O'Riordain & Daniel, 1964, p. 127).

In the remains of the ancient city of Hacilar, in Turkey, the eyes of seated figures were sometimes accentuated through the use of black paint in the pupils. From Eastern Syria, c. 3500 B.C., there are hundreds of figurines of a goddess "whose cult image had staring eyes and eyebrows which met at the top of her nose" (Dames, 1976).

[3]An example of these portable idols is the engraved cow-bone eye idol used to represent the goddess in Neolithic Spain (c. 1500–3000 B.C.). It symbolized the goddess simply by two engraved eyes accentuated by curved lines (Neumann, 1974).

[4]A striking example of this is the shining alabaster Parthian goddess of the luminous moon, from the first or second century A.D. This statue has gleaming, crystalline eyes and navel. The eyes are thus equated with the navel, which Neumann sees as a symbol of "the center from which the universe is nourished" (Neumann, 1974, p. 132).

The symbol of an all-knowing, all-seeing deity belongs also to an archetypical image in which the stars of the night sky appear as the eyes of the goddess (Neumann, 1974). A prehistoric tombstone in female form found in the vicinity of Bologna, Italy, shows the "Queen of Heaven" with two cosmic symbols forming her eyes. "As Goddess of the tomb, she rules over the world of the dead, but at the same time she governs the celestial world, whose luminaries are her eyes" (Neumann, 1974, p. 127).

[5]There can be no direct confirmation that such associations ever entered the minds of ancient artists. However, a group of figures similar to the Warka head, created around 2700–2600 B.C. and found north of Warka in Tel Asmar, indicate that there was indeed an association between vision and supernatural powers. This is shown "by the fact that the two largest figures, identified as deities by the emblems on their bases, also have the largest eyes in relation to their heads" (Gardner, 1980, p. 44).

[6]Eventually, many societies changed from matriarchy to patriarchy and emphasis was placed on gods rather than goddesses (Bachofen, 1967); however, the eye as a symbol of divinity survived.

[7]The all-encompassing vision of the gods also suggests the possibility for the dispensing of true justice. In Assyro-Babylonian tradition, Shamash, a sun god and god of justice, saw everything, and his rays were like a huge net which caught all those who committed wrong (Langdon, 1931).

[8]After the dramatic, public murder of Kitty Genovese, Harlan Ellison was moved to write a story called "The Whimper of Whipped Dogs" (1975) to "explain" the passive complicity of the neighbors who ignored the victim's cries for help. Beth, the story's protagonist, is also a mute witness to a murder in her apartment complex's courtyard. But when she shakes herself from her frozen shock and looks around to other windows, she sees, not horror on the faces of her neighbors, but a kind of bloodlust. And then she looks up above the apartments:

> Beth saw eyes. There, up there, at the ninth floor and higher, two great eyes, as surely as night and the moon, there were *eyes* . . . The eyes, those primal burning eyes, filled with

an abysmal antiquity yet frighteningly bright and anxious like the eyes of a child; eyes filled with tomb depths, ancient and new, chasm-filled, burning, gigantic and deep as an abyss, holding her, compelling her . . . here the eyes of that *other* watched. (pp. 5-16)

The memory of "those eyes" haunts her, drawing her to the window over and over, until the night when she, too, is chosen as the victim of the city, and finds herself on her balcony in the grip of a stranger who has broken into her apartment.

. . . Back, back he bent her further back and she was looking up, toward the ninth floor and higher. . . . *Up there: eyes.*
And then she realized just who the eyes belong to: God! A new God, an ancient God come again with the eyes and hunger of a child. . . . (p. 19)

[9]In Barcelona, Spain, a Romanesque fresco entitled "Apocalyptic Lamb" depicts a reasonably normal animal, except for its seven eyes, each open and alert. Throughout history, the number seven has been accredited with magic power; thus, the number of eyes seems to "endow the creature with extra vision and wisdom" (Kuh, 1971, p. 153).

Stuart Gordon, in his science fiction novel, *Two Eyes* (1975), uses the number of eyes as a major device in this second-coming allegory. When one of the characters (with two eyes) stands before One-eye, the Divine Mutant, his head is filled with images. The "Mutant's eye . . . is like a third eye which completely drowned the perception of his own two eyes, a third eye over which he had no control" (p. 80).

[10]To be stared at in silence by another person makes us feel extremely uncomfortable (Exline, 1971). But we may become especially disturbed or frightened if the eyes are disfigured or colored in an unusual way. After publishing a cover picture of an actor's bloodshot eye, *Smithsonian* magazine received a letter from a reader exclaiming: "EEK! That eyeball on the cover of your March issue! We had to keep the magazine turned over to get that huge red orb 'gazing' anywhere but at us" (April, 1979, p. 10). M. C. Escher (1971) can conjure fear with his "Eye" print, in which a skull appears in the pupil of a wide-open eye.

Science fiction and horror movies very often rely on eyes to create frightening, alien situations. During the credits at the beginning of the 1931 version of *Frankenstein* (Universal Pictures), pictures of eyes revolve in the background. The eyes of many monsters in horror pictures are contorted or disfigured. Some monsters may have blank eyes, some have large eyes, or various numbers of eyes (Naha, 1975). In Michael Bishop's science fiction story, "A Funeral for the Eyes of Fire" (1975), eyes are used over and over to emphasize the strangeness of the aliens, even though their three-lobed heads are at least as odd as their pupilless, lidless, stone-like eyes.

[11]In a French short story called "The Story of the Eye," the horrible image of the eye symbolizes the guilt lovers feel for their sexual activities (Bataile, 1967).

[12]Generations of French children were required to memorize Victor Hugo's poem "La Conscience" (1971), in which the agony of Cain's guilt is

evoked over and over through the image of a watching eye which Cain cannot escape.

[13]The literature of ancient Greece, Babylon, Egypt, and Rome abounds with references to the menacing power of the evil eye. The ancient Greeks had a firm belief in the supernatural power of the eye, which they named "Baskanios" (Evans, 1925). Plutarch asserted that the Thebans could destroy strong adults with a look (Hart, 1949). Perhaps this notion explains why the bows of Greek military ships were often carved with huge staring eyes. The *Athava Beda* and other ancient Indian writings also contain many allusions to the evil eye (Evans, 1925). Belief in its power is mentioned throughout the Jewish Talmud (Werblowsky & Wigoder, 1967), which stated, "For one that dies of natural causes, ninety-nine die of the evil eye" (Baba Metzia, c. 200). Several centuries of Jewish superstition maintained belief in the evil eye, and certain Hebrew expressions accompany compliments in order to protect the receiver from its effects (Trachtenberg, 1974). The Bible also refers to the evil eye, as in the Book of Proverbs: "He that hasteth to be rich hath an evil eye" (28:22).

[14]Although possessors of the evil power may not carry outward signs of their dangerous ability (Evans, 1925), a variety of eye characteristics could make a person suspect: a squint, red eyes, shifty, ill-mannered glances, a heavy sullen look at the ground, a dismembered eye, or eyes that are odd in any way (Evans, 1925; Middleton, 1964).

There are many cross-cultural variations surrounding the belief in the evil eye, as Clarence Maloney (1976) described:

> In one region (the Philippines) the power emanates only from the person who casts it, through the eye or mouth. But in another (the Mediterranean) it may be an avenging, righteous power of a deity, or even (in ancient Egypt) the power of a God emanating from his eyes. The evil eye may be (in Italy) a pervasive malevolent force like a virus or a plague, to be staved off with charms of plastic, or it may be (in Mexico) the effect of bad 'air' over the land, or it may be (among Slovak-Americans) a 'chronic but low grade' evil eye. It may be thought of (in the Mediterranean) as the embodiment of the source of evil, maybe the Devil himself, or even (in the Near East) as a sort of deity in its own right on occasion. But elsewhere (in India) there is no such dichotomy of good and evil rending the world asunder, and the evil eye is only an emanation from the mind behind the eye of the person who casts it. Yet elsewhere (in Guatemala) it is the name of a sickness, not just the cause of the sickness. In the small villages of present day Mexico, if a woman looks too long at a child, the parents will become very angry due to the assumption that the woman is envious and therefore transmitting evil energy through her eyes to the child. Since her glance could result in sickness or death, the child may be taken to a "curanda" or witch who is able to touch the child and thereby cure her/him . . . In the United States too there exist a few people who believe that there are witches and sorcerers possessing the evil eye and can do great damage by intently staring at others. (p. vi)

Many rituals of protection against the evil eye have been developed. At birth, children in Mexico are given a necklace with a deer's eye or facsimile in order to protect them from evil energy transmitted through the eyes. In many Mexican marketplaces and homes one can find statues for protection from such evil influences. Belief in the protective influences of eyes may account for the prevalence of eyes found symbolized in Mexican painting and jewelry. There are eyes on many Mexican necklaces, rings, and earrings.

Protections elsewhere against the evil eye include ritual fumigations with wild rue and the use of charms and amulets (such as miniature Korans, agate, horns, and blue beads, especially ones with spots inside circles, suggestive of irises and pupils). Mirrors have been sewn into clothing to reflect the evil back upon itself. Tattoos of eyes, brightly-colored clothes, and pieces of red cloth have all been used to repel the evil glance. The act of burning alive animals such as cats has been thought to be a form of protection. Holding the two middle fingers down with the thumb, so that the two remaining fingers suggest horns, has been used in attempts to deflect the evil eye. Spitting, recitation of prayers and incantations, and the application of salt have been used as cures of the evil eye, and at times, a less valued possession has had to be sacrificed to protect a more valued possession (Maloney, 1976).

[15]Examination of how eyes are linked with identity in art reveals that the relationship between seeing, being seen, and identity is intensified for visual artists. Because the artistic process requires becoming aware of one's own vision and then representing that vision, it is not surprising to find that artists often use the eye as a symbol of identity in their work. Rembrandt was a master at portraying expression in the eyes. Through his detailed workmanship, the eyes are reflected by radiant waves of light which interrupt dark masses of shadow. Rembrandt used his own eyes as models for his historical paintings (Monneret, 1978). This gives us the impression that Rembrandt could see himself in those historical subjects and that he wanted viewers to see aspects of himself there, also. In Rembrandt's self-portraits of 1652, 1657, and 1660, most of the light in the paintings illuminates the upper face, making the eyes the central focus and the main conveyers of character (Monneret, 1978; Gardner, 1980).

In painter Romaine Brooks' self-portrait, the eyes remain the focal point of the picture, despite the fact that they are in the shadow of her top hat's brim. The observer's eyes may stray to other parts of the image, but they always return to the slightly luminescent and direct gaze of her eyes, which hold you in the shadow. Since this is her self-portrait, and the eyes are such an important part of its effect, the eyes are obviously a symbolic link with her identity.

Artists need to have their work seen in order for the act of creation/ communication to be complete. Thus, unrecognized artists suffer great frustration. It is no wonder that Hans Reichel tried to deal with this problem by painting "gnawing" eyes onto his canvasses in order to "get hold of them (the observers) like a crab, and make them realize that Hans Reichel exists" (Miller, 1960, p. 309). Poet Emily Dickinson, whose life is a "story of genius fashioning its signature in silence and obscurity" (Brinnin, 1960, p. 7), revealed in at least one poem that her sense of self was incomplete because she could not see her work reflected in "Other Eyes" (Dickinson, 1970, p. 192). The use of "Other Eyes" in her poem illustrates the relationship between being seen and a sense of identity. Without other people to read and appreciate her work, Dickinson's sense of self was incomplete.

[17]The association between knowledge and good eyesight, ignorance and defective vision is commonly used in literature. In Cervantes' *Don Quixote*, for example, the Don falls into his misadventures because he does not see clearly. His visual infirmity is used to reflect his "egocentric blindness" (Allen, 1969, p. 89).

In *Bleak House* (1852), Charles Dickens uses the motif of defective vision throughout to indicate the characters' inability to understand the world around them clearly or completely (Ousby, 1977). Mrs. Jellyby, known for her charity toward the suffering of people in foreign lands, overlooks misery at home and exemplifies what Dickens calls, "Telescopic Philanthropy." Dickens describes her as having, "handsome eyes, though they had a curious habit of seeming to look a long way off. As if . . . they could see nothing nearer than Africa!" (Dickens, 1977, p. 37). Because the characters cannot view "the dense crowd of objects" composing their world in a clear or wholistic way, Dickens often has them "peeping," and even the youngest member of the Jellyby family is named "Peepy."

To relate defective vision to ignorance is not the habit of mythmakers and novelists alone. It was a popular belief in the Middle Ages that spectacles (invented circa 1285) were a hood or shield which caused the wearer to misapprehend the truth (Reiss, 1969). Today, people with eye disorders are often avoided or treated with hostility. Persons with deviated visual axes are perceived to be poorly oriented, stupid, dishonest, and of a low socioeconomic class (Lumsden, unpublished paper, n.d.). Heaton (1968) has documented that eye anomalies like an eye patch, a glass eye, and blindness make observers feel uncomfortable. People whose eyes are covered or altered in some way are often judged to be sinister or mysterious.

[18]It is a practice of some who pray or meditate to roll their eyes upwards toward the Third Eye to help awaken spiritual wisdom. Those who meditate may also use mandalas in order to reach heightened spiritual consciousness. Mandalas are Hindu or Buddhist graphic symbols of the universe which are often in the form of a circle divided into sections. There is an unmistakable resemblance between many mandalas and the eyes (Metzner & Leary, 1967; Arguelles & Arguelles, 1972), and because of their use they can be likened to an externalized Third Eye. As the Arguelles (1972) pointed out: "The eye is the human intermediary between the gift of outer light and the light that burns within. According to Matthew, Christ said, 'The eye is the light of the body; if, therefore, thine eye be single, thy whole body be full of light.' The mandala can be described as a symbol or vehicle of the process described by Christ whereby the eye is made *single*, like a lens, flooding the organism with light—not the light of the sun but the inner light of which the sun seems but a reflection" (p. 23).

# APPENDIX I

# Measuring Eye Contact

In order to assess the relative significance of eye contact in a particular encounter, methods for measuring it and determining when eye contact occurs have been developed. This need for accurate measurement has posed a variety of methodological problems for researchers.

Terms such as "visual interaction," "eye contact," "mutual gaze," "eye signals," and "eye behavior" have been used interchangeably and without much qualification in our discussion of eye behavior thus far. This interchangeability was due to the general nature of references to the role of eyes and looking behavior between persons. However, in order to reach a more precise understanding of how our eyes operate in social interactions, it is now important for researchers to discriminate as much as possible between types of looking behavior and to isolate the eye behavior in which they are most interested.

Researchers (e.g., Harper et al., 1978) have bemoaned the "vagueness and apparent arbitrariness" (Kirkland & Lewis, 1976) of the terminology used to describe visual interaction in the psychological literature. In 1968, for example, Mehrabian (1968a) operationally defined eye contact as the subject's head orientation toward an imaginary other person with whom the subject was instructed to engage in conversation. In addition to the fact that judgment of the subject's head orientation was imprecise, Mehrabian's experimental definition of eye contact did not make logical sense: "eye contact" implies a mutual activity, and in Mehrabian's experimental situation, it could not be known when the imaginary partner was returning the glance.

In 1971, von Cranach did much to clarify the vocabulary of visual behavior by offering definitions which differentiated the reciprocated from the unreciprocated look. His definitions also illustrated the variety of looking behaviors which are possible in visual interactions: one may direct one's gaze in the direction of another's face ("face gaze") or in the direction of another's eyes ("eye gaze"), and if these gazes are not returned, they are "one-sided looks." Two persons may gaze at each other's faces simultaneously in a "mutual look," or they may look into each other's eyes with awareness of each other's gaze and have "eye contact." "Gaze avoidance" occurs when one evades another's gaze; "gaze omission" occurs when one fails to look at the other without the intention of avoiding eye contact. Tankard (1970) added another level of precision to a definition of eye contact: it can be said to occur between two persons "when the image of the eyes each falls on the fovea of the other's eyes. Eye contact is broken when one person moves his eyes so the image of the other person's eyes is no longer on his fovea" (p. 49).

For our purposes, eye contact will be defined as occurring when two persons each look simultaneously at the eyes of the other. This phenomenon has also been named "mutual glance" (Simmel, 1921), "interocular intimacy" (Tomkins, 1963), "eye-to-eye looks" (Goffman, 1963), "looking into the line of regard" (Lambert & Lambert, 1964), "mutual or shared glances" (Exline, 1963), "mutual gaze" (Argyle & Ingham, 1972; Fehr & Exline, in press), and "eye engagement" (Exline & Fehr, 1978).

Like other forms of nonverbal communication, eye contact constitutes a "signal" between persons (Exline & Fehr, 1978). In order to verify and identify the signaling functions of eye contact, it is necessary for experimenters to demonstrate that this signal is indeed perceived by a recipient. Thus, it has become important for investigators to obtain a valid measure of eye contact in their research. Much controversy has surrounded this issue. While some investigators feel that people can tell when they are being looked at in the eyes, others feel that this ability is quite limited (Exline & Fehr, 1978).

Most of us who are sighted assume that we have the ability to observe and measure fairly accurately the presence or absence of the visual attention of those with whom we interact or of those we observe interacting. We believe that we know when we are being looked at, or shunned, because of the salient, arousing, and involving characteristics of the eyes. Many investigators of eye contact also have assumed that the signal of one person looking into the other's eyes could be accurately detected by the receiver. Some authors have justified this assumption on the basis of a study by Gibson and Pick (1963), who investigated the ability of observers to determine whether they were being looked at in the face. They found that the observers' judgments were based somewhat on the position of the sender's head but mostly on that of the eyes. They concluded that the "ability to read the eyes seems to be as good as the ability to read the fine print on an acuity chart" (p. 394).

Cline (1967) replicated Gibson and Pick's experiment using more refined methods. He qualified their results, finding that eye and head position interact, causing perceived gaze direction to fall between the positions of the head and eyes. When the head position remains stable and directed straight ahead, and gaze direction changes, the perception of whether a person is looking is determined by the eyes' position. Cline, as well as Anstis, Mayhew, and Morley (1969), noted that when the line of regard is directed into the receiver's eyes it is more accurately perceived than other lines of regard.

On the other hand, von Cranach and Ellgring (1973) found that although the receiver of the glance could usually judge whether the look was directed at his/her face, s/he could not accurately judge eye contact. Von Cranach and Ellgring (1973) also tested the ability of a distant observer (who was not involved in the sender-receiver interaction) to perceive looking behavior, thus approximating the methodology used in a large number of studies on eye contact. Von Cranach and Ellgring (1973) cast doubt on the results of such studies when they found that an observer who is farther from the sender than the receiver differentiates considerably less accurately than the receiver. Stephenson and Rutter (1970) also found that as distance between interactants increases, there is a greater chance that observers will mistake gazes directed near the eyes (e.g., at the ears) as eye contact. Von Cranach (1971) found that head and body position considerably influence the observer's judgments.

Knight, Langmeyer, and Lundgren (1973) examined similar methodological issues associated with standard procedures of studying eye contact. They concluded that the common techniques of measuring eye contact were inadequate because the observer's expectations influenced their perception of how much eye contact occurred. They proposed the use of alternative techniques for measuring eye contact; for example, video recordings of interactions using a zoom lens. Observers not only bring different expectations to the task of judging eye contact, but also different skills and styles, all of which may be influenced by the individual's cultural background, gender, and age. Thus, as Woolfolk (1981) pointed out, procedures and precautions are essential for establishing observer reliability, as well as frequent retraining, practice, and spot-checking of observers throughout the course of an experiment.

In many experiments which measure eye contact as a dependent variable, the confederates either fix their gaze on the subject's eyes throughout the entire observation period or for predetermined intervals (Harper et al., 1978). Such contrived, unnatural situations have been found to confound the results of these experiments. For example, Kendon (1967) showed that, in an interacting dyad, the visual behavior of each person is significantly affected by the visual behavior of the other. Fischer (1972) found that confederates felt uncomfortable when they were required to look ten inches to the side of the listener. This discomfort was apparently communicated unintentionally to persons who viewed a video transmission of the interaction in which the eyes of the confederate were masked out. Although the viewers could not see any difference in visual behavior between the eye-contact/no-eye-contact conditions (because the eyes were blocked from view), they showed a significant preference for the confederate in the eye contact condition over the confederate in the "look ten inches to the side" condition. Apparently, the confederate speakers were unable to hold other nonverbal cues constant when shifting gaze direction and these unintentional cues significantly affected the preference decisions of the study's participants. Other aspects of experimental situations can also affect perception of another's gaze, e.g., visual acuity, lighting, observer training, and whether the behavior is "in person" or on a TV screen (Exline & Fehr, 1978).

When involved in everyday focused interactions with other persons, we usually turn our eyes and head toward the other's face. The occurrence of such "face reactions" has led Dittman (1978) to assert that one of the best ways to identify eye contact in the research situation is to note the accompanying head movements: If an individual has turned her/his head to face another's, they are likely to be looking into each other's eyes. For Dittman, then, eye contact does not need to be defined precisely—the turn of one's head to gaze at another's face is close enough. Vine (1971) also suggested that everyday, spontaneous eye interactions are more easily perceivable than "programmed" laboratory interactions because people tend either to look directly at another's eye region or to look away. In a comprehensive review of the experimental literature on social, visual interaction, Fehr and Exline (in press) noted that observers' and receivers' measurements of "on-face and off-face gazes" have been found to be reliably similar.

Obviously, studying human eye behavior in an experimental situation has its drawbacks. In their attempts to study eye contact, investigators have removed their subjects from normal, everyday contexts, thereby allowing systematic observation, but casting doubt on the applicability of results. Perhaps a technology will develop which can control the experimental conditions that affect observers' abilities to perceive eye contact. Then we may know more accurately how well humans can distinguish the occurrence of eye contact. Given the inherent importance of visual behavior among humans (as indicated by subjective experience and a growing body of research on humans and primates), it seems reasonable to assume that the human capability to detect accurately the direction of the look will be experimentally verified in the future.

# APPENDIX II

# The Human Visual System

We must not limit our study of the visual communication system to only the study of visual signs and their behavioral results. To do so would be to deny the great complexity of the human creature whose "behavior can be understood only if the biophysical and symbolic processes are encompassed in one over-all system" (Ruesch, 1972, p. 12). Ruesch, a leader in the attempt to create an interdisciplinary, holistic approach to human communication, asserted that "a multivariate approach to the understanding of human affairs must contain the following: the inside view—psychology and psychoanalysis; the outside view—anthropology and sociology; and the component parts analysis—biological and medical science" (1972, pp. 39–40). Ruesch encourages us to disregard the usual mind–body dichotomy which splits biological/physical knowledge from psychological/social knowledge. In the interest of bridging such dichotomous approaches and because there is an "intimate relationship" between anatomical form and function (Chevalier-Skolnikoff, 1973), the biological apparatus involved in visual communication is described in this appendix.

The visual system in humans consists of the eyeballs; the muscles which move the eyeballs within rounded skull cavities (known as the orbits); the protective eyelids; the optic nerve, which connects the eyes to the brain; and the brain, which perceives the visual images and decides the response the body will make to visual information.

The human eyeball is capable of free movement in every direction except laterally. We can look upward easily, whereas most animals need to use the neck and lift the head to do so. Six muscles, four straight and two oblique, are connected to the eyeball and allow its movement, as Smythe (1975) described: "The straight muscles (recti) pull the eyes towards their respective sides. The superior oblique draws the cornea up and rotates the eye outward" (p. 46). The correct functioning of the muscular system is important for normal vision. A failure in muscular functioning can cause double vision and other abnormalities (Smythe, 1975).

The mobile capabilities of the eyes are cited by E. B. Forrest (1969) as indicative of the dynamic nature of the human visual system. As Forrest observed, it is a mistake simply to consider the eyes as passive sense receptors for "the only major sensory aspect of the eye is the retina" (1969, p. 38). The eyeball is more than just an optical sensory system. It is also "one of the most complex motor organs in the body" (Forrest, 1969, p. 38). The eyeball is able to adjust itself to control light distribution on the retina. It reacts to this light distribution with one of three possible responses: following the stimulus; shifting to follow another stimulus; or avoiding a given stimulus completely. With its unique patterns of movement, the eyeball motor-sensory system aligns itself onto a target with one muscle group and focuses in on that target with another muscle group (Forrest, 1969).

Within the eyeball the electromagnetic radiation we call light is transformed by photosensitive receptor cells into energy to which the brain can respond (Isaacson, Douglas, Lubar, and Schmaltz, 1971). The eyeball is an almost perfect sphere except for a protruding, curved portion which forms a transparent "windowpane" at the front of the eyeball's protective coating (Sutton-Vane, 1958). This transparent, protruding section is the cornea. The rest of the protective coat surrounding the eyeball and maintaining its shape is known as the sclera, which we see as the "white" of the eye. The sclera is firm and fibrous and covered by a thin moist skin called the conjunctiva (Smythe, 1975; Sutton-Vane, 1958). Beneath the sclera lies the black or dark brown choroid coat which, through its deep pigmentation, prevents loss of light or the scattering of light rays (Smythe, 1975; Sutton-Vane, 1958).

A little more than half the eyeball resides within the bony cavity of the orbit. The rest extends outside the orbit, but is well protected by large, mobile upper and lower eyelids. The eyelids carry lashes which keep dust and grit from falling on and damaging the very sensitive cornea (Smythe, 1975).

Light enters the eye through the cornea and then passes through a pocket of salty liquid, the aqueous humor, which nourishes the cornea and the lens. Between the cornea and the lens is the iris, which is usually pigmented (albinos are the exception). The iris is a "membranous structure containing both circular and radial muscular fibres" (Smythe, 1975, p. 34) with an opening in its center called the pupil. The iris acts as a "window shade" (Sutton-Vane, 1958, p. 14) which responds automatically to the amount of available light. When the light is low, the radial muscular fibers in the iris contract and increase the size of the pupil. When the light is bright, the circular muscular fibers act to decrease the iris opening. The pupil also contracts and dilates in response to our emotional states.

Once through the pupil, light rays move through the soft, crystalline lens. Slightly convex in front and markedly so behind, the lens focuses light on the retina which lies on the back and sides of the eye's inner walls. A suspensory ligament holds the lens in place behind the iris. This ligament is attached to ciliary muscle which, when contracted, pulls on the ligament, which in turn pulls the lens, thus causing the elastic lens to flatten. This process, called accommodation, allows the lens to adjust for sharper focus depending upon the distance of the objects to be seen (Smythe, 1975).

In order for light to find its way to the retina, it must pass through the vitreous humor, a clear, jelly-like substance which fills out the eyeball behind the lens (Sutton-Vane, 1958), about four-fifths of the total volume of the eye (Smythe, 1975). Finally, light reaches the retina, which some say is like a part of the brain itself (e.g., Smythe, 1975). "It is by far the most sensitive and highly developed part of the whole eyeball . . ." (Smythe, 1975, p. 52). It is here that the actual transformation of light energy into nerve cell energy takes place. The entering waves of light activate the photosensitive receptor cells (rods and cones) which transmit their energy to the bipolar and then the ganglion cells which lie in the retinal layers above (Isaacson et al., 1971). The ganglion cells eventually gather into a "cable" which leaves the eye as the optic nerve, transmitting the impulses sent from the rods and cones to the visual cortex and other areas of the brain where light patterns are then "perceived" (Isaacson et al., 1971; Sutton-Vane, 1958).

The area in the retina where the optic nerve leaves the eye contains no rods or cones. This is known as the "blind spot" because it cannot register any part of a visual image. Although we are not conscious of this "blind spot," its existence can be demonstrated, as Smythe (1975) described: "A simple method is to hold a forefinger erect and about eighteen inches in front of your face. Close one eye and keep the other firmly fixed upon an object in front, such as a light bulb. Now keeping at the same distance from the face, move the finger and arm slowly to one side. You will find one position in which you can no longer see your finger, but on moving it a little farther, it will again appear. The point of disappearance corresponds with your blind spot" (p. 51).

Rods are adapted to night vision; the cones are adapted to daylight and color vision. There are about four times as many rods as cones in the retina, with a large number of rods in the peripheral areas of the retina. The cones are more plentiful in the central (macula) area (Isaacson et al., 1971). In the center of the yellow, oval macula lies the fovea centralis, which contains only cones and is the point of most acute daytime vision. It is upon the fovea that the visual image is most sharply focused (Isaacson et al., 1971; Sutton-Vane, 1958). At night, however, it is wise to look a bit to the side of where you expect to see something (such as a star or approaching light) since only the rods are sensitive to low illumination and they are most plentiful on the peripheral areas of the retina.

The complex human nervous system, the brain in particular, allows us to "integrate, synthesize, alter, and get meaning from the image placed" (Forrest, 1969, p. 38) on the retina. McCulloch (1951) said that ". . . of the three million inputs the brain can receive per millisecond, two million are visually dispatched" (cited in Forrest, 1969, p. 39). Hall (1963) has described neurophysiological evidence indicating that the eyes send from six to twenty times as much input to the brain as the ears. With their dynamic capabilities and large neural input, the eyes have "the potential for processing the most useful information; especially in terms of awareness, knowledge, and action" (Forrest, 1969, p. 39). The unique information-processing potential of the eye leads Forrest to suggest that vision may be the "dominant" sense (1969, p. 39).[1]

## NOTE

[1]We must be careful, however, not to isolate vision from the totality of interrelated sense systems of the body. It is true that for sighted persons our awareness of the external world is based mainly on information received from vision, but hearing and smell also play a role in consciousness of things distant from us. This distance information is then "combined with proximal information from touch and taste and information from internal systems that tell us where our head and limbs are and something about many other internal states" (Jerison, 1973, p. 18). The relatedness of the different sense modalities is further illustrated by the fact that "active tactual exploration with the limbs is crucial for visual learning in mammals" (Kruger & Stein, 1973, p. 82).

# APPENDIX III

# *Visual Evolution*

The human eye is representative of the mammalian eye in general, but there are many variations among species. We are often able to gain insights into the physiology and behavior of human beings by studying those of animals, and many believe there is an evolutionary relationship between humankind and other inhabitants of the earth. Indeed, Paul Ekman (1973) has reminded us that "Darwin (1872) claimed that we cannot understand human emotional expression without understanding the emotional expressions of animals, for, he argued, our emotional expressions are in a large part determined by our evolution" (p. ix). Thus, evolution of visual emotional expression and other functions of looking behavior cannot be separated from the evolutionary development of the anatomy of the visual system.

Darwin explained that the apparent kinship among different forms of life on earth is due to the fact that we all evolved from a common ancestry (Meuler & Rudolph, 1966). Evolution from one form to another takes place, he said, through the mechanism of natural selection or "survival of the fittest," a process which takes thousands of years. The concept of natural selection would explain that spontaneously occurring changes or mutations in the visual systems of certain animals contributed to their survival and thus, the reproduction of their kind. Animals which "retained or developed eyes of inferior quality soon became the victims of other animals and failed to reproduce, with a result that the inferior eyes disappeared and a more useful variety took their place" (Smythe, 1975, p. 13).

The survival of any animal depends upon the ability of its sense organs to respond appropriately to its unique environment and in accordance with its specific needs and limitations. Thus, visual progress in one species may mean progress in quite a different direction for another. As Smythe (1975) pointed out, "The eye of a bird would be as useless to a rat as those of the rat to the bird" (p. x). Whether an animal spends most of its time on the ground, in the air, under the water, or on the water's surface, and whether it is active in bright light or darkness, determines the specialized visual characteristics a species requires.

According to Kruger and Stein (1973), the varying visual needs among species make it questionable to search for a visual system which represents the original, primordial system from which all others evolved. To describe an invertebrate system as "primitive," for example, may lead us to mistakenly believe that it is a "step in the development of a more 'advanced' system, rather than being an example of unique sensory specialization" (Kruger & Stein, 1973, p. 65). With this in mind, the usual hierarchical framework employed by evolutionary theory—in which some species are called "higher" or more "advanced" than others—appears quite arbitrary. Kruger and Stein (1973) illustrated their point with this example: "The impressive 'specialization' in man is expansion in the size and 'complexity' of the microstructure of his brain, but can one satisfactorily place this 'higher' than the small brain of the honeybee which is capable of enormous behavioral complexity with far fewer neurons" (p. 64)? Kruger and Stein concluded that the concept of an "advanced" species is, essentially, based upon human value judgment (which reflects our species chauvinism).

Other biologists today are also questioning orthodox Darwinism (Davy, 1981). For example, Maynard Smith of Sussex University pointed out that if the main problem for life is survival, then simple "primitive" organisms are a fine solution—why should there have been changes toward increased complexity (Davy, 1981)? Dialogue on such issues is in its infancy, but the reader should be advised that ideas presented here on the evolutionary development of visual systems and behavior may be called into question in the future.

As far as human beings are concerned, our eyes are neither better nor worse than they were two thousand years ago (Smythe, 1975), yet our technology has introduced many new kinds of visual experiences. Smythe (1975) has implied that human eyes have not evolved to keep up with our technological development. For example, he cited the inability of the eyes to function adequately under high speed driving conditions as the reason why "more human lives throughout the past fifty years (1925–1975) have been lost [in automobile accidents] than in all the wars in history" (1975, p. 6).

In a mammoth study of *The Vertebrate Visual System*, Polyak (1957) noted that the eye is among the very first organs to appear in the human embryo. Polyak considered human embryonic development to be a record of the stages of the evolution of human ancestors, a view accepted by other scientists (e.g., Haeckel, 1900; Dodson & Dodson, 1976). The early emergence of the eye in ontogenetic development thus indicated to Polyak that photosensitive animal substance represented the matrix from which not only the eye, but vertebrates in general, developed. This point illustrates the basic relationship between solar light and life on earth. Indeed, Polyak believed light may have been an initiator of life,[1] and he asserted that, "It is upon light that animals and plants alike, in the final analysis, depend for their origin and existence" (1957, p. 1046).

The relationship between most of earth's life forms and light energy "can easily be seen in the turning of leaves and flowers toward the sun and in the movement of many small animals, which possess no true eyes, towards or away from light" (Mann & Pirie, 1962, p. 10). Even amoebae, one-celled animals with no true sense organs, display this movement, which is called "phototropism" or "heliotropism." This ability of light to fall upon a cell, change it, and make it do something is the basis of the visual process.

A dermal (skin) light sense has been found in many forms of life. For example, in reaction to changes in light, coelenterates (jellyfishes, sea anemones, corals, etc.) exhibit withdrawal reactions; cyclostomes (e.g., the eel-like lamprey) and blind fish respond by moving, and sea urchins exhibit localized covering responses. In some multicellular animal organisms, certain specialized cells (rather than the whole body) are chemically affected by light energy, as in the earthworm whose light-sensitive cells are indiscriminately scattered over the skin (Mann & Pirie, 1962). The changes in skin color which occur in teleosts, amphibians, and many reptiles are due to a rearrangement of melanin granules within certain skin cells when light enters them (Kruger & Stein, 1973).

In some sea-living invertebrates like the marine worm, light-sensitive cells are gathered into a little patch which is spread out flat on the surface (Mann & Pirie, 1962). The "eyespots" of such creatures cannot see us unless we move, causing illumination on one side of the spot to decrease as illumination on the other increases. Usually, the perception of movement through the eyespot causes the animal to move reflexively, as in the animals mentioned above. This reaction is a defense mechanism which adds to the chances of the animal's survival. They "move first and think later. They live longer that way" (Smythe, 1975, p. viii).

There are many markings which resemble eyes on the euglena, a minute, protoplasmic life form. There is only one actual light-sensitive eyespot, however, which is a black patch near the base of the flagellum (hairlike appendages). A microscope reveals that this patch is a rose-colored round protuberance made up of six small spots or lenses. The eye of the early euglena would get injured constantly because it stood out beyond the skin's surface. Sutton-Vane (1958) proposed that this is why "the eyes of some Euglenas grew less prominent, and in time the rosy little boss of magical pigment became habitually set in a cuplike depression" (p. 30). Dark pigmentation probably developed in eyespots because it prevented undue loss of light and also provided protection against exposure to ultraviolet rays and resultant "sunburn" (Sutton-Vane, 1958).

Life in the sea has flowered over thousands of centuries into a variety of multicellular invertebrates. Varieties of light-sensitive cells and neural apparatus are accompanied by numerous evolutionary solutions to the problem of forming an image. For example, in the visual pit of the limpet (a mollusk), the light-sensitive cells are concentrated in a small depression covered by a secretion which increases intensity and the contrast of shadows, but which does not operate as an image-forming lens. (In order for an image to be formed, rays of light from surrounding objects must be *focused* on photoreceptive cells.) In the eye of the Nautilus (also a mollusk), there is a very small opening, the size of a pinhole, which serves to enhance formation of an image, as well as to protect the photoreceptor cells.

The unique pinhead-sized marine animal *Copilia*, which some scientists believe may be an evolutionary link between eyespots and more complex eyes, can actually form images through a crude lens system (Mueller & Rudolph, 1966). Two lenses on one side of the *Copilia* reflect light to a smaller set of lenses deep within its transparent body. The interior lenses are each set on hook-shaped light receptors which scan back and forth while sending signals to the brain. The *Copilia* is the smallest creature capable of perceiving images (Meuller & Rudolph, 1966).

The scallop has exceptionally good vision for a mollusk because it has from fifty to one hundred very small eyes studding the fleshy mantle just inside its shell. Each eye has its own lens and a group of photoreceptor cells. The scallop has a greater need for good vision because, unlike clams or oysters, it can propel itself across the ocean floor (Mueller & Rudolph, 1966).

The eyes of the cuttlefish and octopus, with their cornea, lens, vitreous, iris, retina, and other detailed structures, closely resemble vertebrate eyes (of which the human eye is a representative example). Polyak (1957) said that the appearance of such chamber eyes "equipped with a dioptrical (refractive lens) apparatus must be looked upon as the next important phase in the evolution of visual mechanisms after the differentiation of the photoepithelium (light sensitive cells)" (p. 786).

Paired eyes first appeared in the cephalopods (e.g., the nautilus, squids, octopuses), according to Sutton-Vane (1958). They give the animal the advantage of being able to orient itself in reference to its "own longitudinal mid-plane and to the surface of the ocean" (Sutton-Vane, 1958, p. 33). Sense of perspective is also enhanced by the overlapping, superimposed fields of vision of the two eyes. Because each eye sees the visual field at a slightly different angle, the two images received are not exactly alike. The brain, however, perceives the two images as one and interprets the differences as being due to the varying distances of the objects from the eyes. This capacity for a sense of perspective or depth is called stereopsis (Mann & Pirie, 1962).

Those who would trace our visual ancestry to other species disagree as to who our ultimate visual ancestor may have been. Mann & Pirie (1962) stated that the phototropism of the amoebae was the first step toward sight, a view also put forward by Sutton-Vane (1958) in her discussion of primitive flagellata. Kruger and Stein (1973) cited evidence indicating that dermal light sensitivity may be related through evolution to the original *touch* neural systems, and not necessarily only to vision. Smythe (1975) stated that "the first eye possessing characteristics comparable with modern eyes probably appeared in the amphibious age, when animals made use of them on land and under water. Later they became adapted to use on dry ground, when our ancestors found it easier (i.e., most beneficial to survival) to crawl out of the water and enjoy the more plentiful produce of the land" (p. 8).

Indeed, some scientists have gone so far as to say that the evolution of the vertebrate eye was the most important development for enabling humankind's distant ancestors to rise out of the sea and survive on dry land (Mueller & Rudolph, 1966).

In *The Evolution of Brain and Intelligence* (1973), H. Jerison estimated that reptiles, on the next evolutionary level beyond amphibians, adapted to living in niches on the land more than 200 million years ago. Jerison (1973) assumed that visual information about events happening at a distance had survival value for the reptiles. That assumption, coupled with the contemporary observation of reptilian eyes, led Jerison to conclude that "early reptiles lived in a visual world . . . The extensive neural information-processing machinery of their retinas would have enabled them to respond selectively to many different stimuli" (1973, p. 18).

Reptiles exhibit the general nonaquatic, vertebrate pattern of visual systems in which corneal refraction, variable lens curvature, and variable aperture (provided by the iris/pupil mechanism) result in "exquisite-image resolution on the retinal photoreceptors" (Kruger & Stein, 1973, p. 71). Reptilian visual behavior is extremely stereotyped, however, due to the limited brain development (Jerison, 1973). A model of visual stereotypy is the response of frogs (which are amphibious) to moving stimuli in Sperry's (1951) experimental situation. Sperry permanently rotated a frog's eye in its socket through surgical intervention. This caused the frog to perceive inverted images, to which it never adjusted. Instead, it persisted in striking out at the point of stimulation as it would have been mirrored on the retina of a normal eye (Sperry, 1951).

The perceptual space of people can be inverted through the use of special lenses (Kohler, 1963). Unlike frogs, Kohler's (1963) subjects were able to modify their behavior according to their experience. People perceived the inverted image from the Kohler lens eventually to right itself after persistent exposure. Evolution has apparently resulted in the reduction of visual stereotypy in human beings.

Whereas the main adaptation of reptilian life forms had taken place in the eyeball, the next major evolutionary turning point occurred in neural apparatus. At the opening of the "age of mammals" about 60 million years ago, selection pressures resulted in the appearance of more extensive neural networks in the mammalian brain, especially among primates.

The type of brain to which a set of eyes is connected is crucial to the amount of use which an animal can make of its eyes and the amount of control it has over them (Mann & Pirie, 1962). For example, the "mechanically perfect" eyes of the seagull would allow it to see the finest details of the smallest print, but it could not, of course, read (Mann & Pirie, 1962). Humans do not have the best eyes, but they have the "greatest consciousness and the biggest brain and therefore the greatest possible number of alternative actions in response to any given stimulus . . ." (Mann & Pirie, 1962, p. 19).

Basic to all vertebrates is the reflex arc of impulses which pass from sensory nerve fibers to the spinal cord and back through motor nerves to the muscles. This allows automatic, unconsciously-motivated, reflexive behavior (Chevalier-Skolnikoff, 1973). The next higher level of neural complexity includes a hindbrain and hypothalamus which allow the mediation of stereotyped behaviors concerned with vital functions of the organism, such as sexual or emotional behavior (including facial expression) (Chevalier-Skolnikoff, 1973).

When brain size increased to form a cerebral cortex with areas of mediation for each sense organ system, it was an "extraordinary" evolutionary advance (Kruger & Stein, 1973). The cerebral cortex receives stimuli from the environment, mediates voluntary, planned behavior and also complex, learned behavior (Chevalier-Skolnikoff, 1973). Within the cortex there are variations among the amount of neural tissue associated with the different sense functions. The more "visual" a species is, the more enlarged will be those portions of the brain which mediate vision (Jerison, 1973). In the more "advanced" primates, the size and perhaps the number of cortical fields concerned with vision are larger than those of other mammals (Kruger & Stein, 1973). Animals with an especially large cerebral cortex are humans, apes, elephants, whales, and dolphins.

Overall, a fairly standard eye serves all vertebrate species. But differing environmental demands have caused the appearance of certain unique anatomic and biochemical adaptations (Kruger & Stein, 1973). Whether an animal is most active in the night world (nocturnal) or during the light of day (diurnal) has caused some of the major differences in eye adaptations among vertebrates. Nocturnal animals possess predominately rod retinas, while diurnal species have varying proportions of cones present, as well as rods (Kruger & Stein, 1973).

Over millenia of evolution, animals who inhabit gloomy caves, underground burrows, and the world of night have developed either very large eyes or have sharpened other senses to compensate for weak, ineffective eyes (Mueller & Rudolph, 1966). The night-prowling owl, for example, has eyes so large they cannot turn in their orbits. The mole rat, with eyes no bigger than pinheads, is almost blind and can only detect shades of light and dark, which is useful in perceiving if its burrow has been broken into from above (Mueller & Rudolph, 1966).

The prosimian, "the most primitive primate" (Chevalier-Skolnikoff, 1973), has not developed the visual capacities typical of other Old World monkeys because it has remained nocturnal, and relies primarily on smell, sound, and touch for information about its environment. A dramatic shift toward visual functioning occurred in other Old World monkeys when they adapted to a diurnal, tree-living life. In order to leap rapidly through the trees, keen vision became a requirement (Chevalier-Skolnikoff, 1973). Visual communication is important for more earth-bound monkeys, such as the macaques, and for the savanna (plains) species, such as the baboons and patas monkeys. Vocal communication might attract predators, but visual communication can be easily and surreptitiously perceived in open grasslands (Chevalier-Skolnikoff, 1973).

Along with the increased visual capacities developed by the Old World monkeys came an increase in the size and number of midfacial muscles (Chevalier-Skolnikoff, 1973). These muscles permitted a new range of eye and facial expressiveness which may have increased survival opportunities for monkeys who live in tightly knit social groups, as most Old World monkeys and apes do. The close proximity of group members makes the use of visual signals useful and feasible (Chevalier-Skolnikoff, 1973). Thus, an animal's social, as well as physical, environment creates selection pressures for visual adaptation (Marler, 1965).

Facial markings in the form of hair patterns or skin color have also been selected in the course of evolution to accentuate eye expressions in some primates. For example, in the mangaby monkey the upper eyelids (which are lowered to threaten other animals) are colored white, which contrasts greatly with the rest of the face. This creates the appearance of huge white eyes from a distance (Chevalier-Skolnikoff, 1973).

As the visual evolution of monkeys illustrates, the selection pressures of physical and social environments contributed to the development of certain anatomical structures. Natural selection also probably contributed to the development of various modes of visual communication behavior as well.

## NOTE

[1] For a detailed understanding of the possible origins of life on earth and the role light may have played, see Alexander I. Oparin's *The Chemical Origin of Life*. Springfield, IL: Charles C Thomas, 1964.

# References

Ackerman, F. E. (1979). Personal communication, May 19.

Agel, J. (Ed.). (1973). *Rough times.* New York: Ballantine Books.

Ahrens, R. (1954). Beitrag zur entwicklung des physiognomie und mimikerkennens. *Zeitschrift fur Experimentelle und Angewandte Psychologie, 2,* 412–454.

Aiello, J. R. (1972a). Male and female visual behavior as a function of distance and duration of an interviewer's direct gaze: Equilibrium theory revisited. (Doctoral dissertation, Michigan State University, 1972.) *Dissertation Abstracts International, 33/09B,* 4482B–4483B. (University Microfilms No. 73-5312.)

Aiello, J. R. (1972b). A test of equilibrium theory: Visual interaction in relation to orientation, distance and sex of interactants. *Psychonomic Science, 27,* 335–336.

Alexander, A. (1979, May 20). A commune's last stand in the Tennessee Hill Country. *Washington Post,* pp. H1, 4.

Alexander, F. (1961). *The scope of psychoanalysis.* New York: Basic Books.

Allen, D. E., & Guy, R. F. (1974). *Conversational Analysis.* The Hague, Netherlands: Morton.

Allen, D. E., & Guy, R. (1977). Ocular breaks and verbal output. *Sociometry, 40(1),* 90–96.

Allen, J. (1969). *Don Quixote: Hero or fool?* Gainesville: University of Florida Press.

Altman, J. H. (1971). Identification of personality traits distinguishing depression-prone from non-depression prone individuals. (Doctoral dissertation, Rutgers University.) *Dissertation Abstracts International, 32/02B,* 1202. (University Microfilms No. 71-20041.)

Altman, S. A. (1967). The structure of primate social communication. In S. A. Altman (Ed.), *Social communication among primates* (pp. 325–362). Chicago: University of Chicago Press.

Amalfitano, G., & Kalt, N. C. (1977). Effects of eye contact on evaluation of job applicants. *Journal of Employment Counseling, 14,* 46–48.

Ambrose, J. A. (1961). The development of the smiling response in early infancy. In B. M. Foss (Ed.), *Determinants of infant behavior* (Vol. 1, pp. 179–196). New York: John Wiley.

Amerikaner, M. (1980). Self-disclosure: A study of verbal and coverbal intimacy. *Journal of Psychology, 104,* 221–231.

And now a word about commercials. (1968, July 12). *Time Magazine,* pp. 55–60.

Andrew, R. J. (1963). The origin and evolution of the calls and facial expressions of the primates. *Behavior, 20,* 1–109.

Anstis, S., Mayhew, J., & Morley, T. (1969). The perception of where a face or television single "portrait" is looking. *American Journal of Psychology, 82,* 474–489.

Appel, V. H., McCarron, L. T., & Manning, B. A. (1968). Eyeblink rate: Behavioral index of threat? *Journal of Counseling Psychology, 15,* 153–157.

Arguelles, J., & Arguelles, M. (1972). *Mandala.* Berkeley, CA, and London: Shambhala.

Argyle, M. (1968). *The psychology of interpersonal behavior.* Baltimore: Penguin Books.

Argyle, M. (1969). *Social interaction.* London: Methuen.

Argyle, M. (1970). Eye contact and distance: A reply to Stephenson and Rutter. *British Journal of Psychology, 61,* 395–396.

Argyle, M. (1973). Eye contact. In D. C. Mortenson (Ed.), *Basic readings in communication theory.* New York: Harper & Row.

Argyle, M. (1975). *Bodily communication.* London: Methuen.

Argyle, M., & Cook, M. (1976). *Gaze and mutual gaze.* Cambridge: Cambridge University Press.

Argyle, M., & Dean, J. (1965). Eye contact, distance, and affiliation. *Sociometry, 28,* 289–304.

Argyle, M., & Ingham, R. (1972). Gaze, mutual gaze and proximity. *Semiotica, 6,* 32–49.

Argyle, M., Ingham, R., Alkema, F., & McCallin, M. (1973). The different functions of gaze. *Semiotica, 7,* 19–32.

Argyle, M., & Kendon, A. (1967). The experimental analysis of social performance. In L. Berkowitz (Ed.), *Advances in Experimental Social Psychology* (Vol. 3, pp. 55–98). New York: Academic Press.

Argyle, M., Lalljee, M., & Cook, M. (1968). The effects of visibility on interaction in a dyad. *Human Relations, 21,* 3–17.

Argyle, M., Lefebre, L., & Cook, M. (1974). The meaning of five patterns of gaze. *European Journal of Social Psychology, 4,* 125–136.

Argyle, M., Salter, U., Nicholson, H., Williams, M., & Burgess, P. (1979). The communication of inferior and superior attitudes by verbal and nonverbal signals. *British Journal of Social Clinical Psychology, 9,* 222–231.

Argyle, M., & Williams, M. (1969). Observer or observed? A reversible perspective in person perception. *Sociometry, 32,* 396–412.

Ashear, V., & Snortum, J. R. (1971). Eye contact in children as a function of age, social and intellective variables. *Developmental Psychology, 4,* 479.

Bachofen, J. J. (1967). *Myth, religion and mother right.* Princeton: Princeton University Press.

Baker, C. (1976). Eye openers in ASL. *California Linguistics Association conference proceedings.* San Diego: San Diego State University.

Baker, C. (1977). Regulators and turn-taking in American Sign Language discourse. In L. A. Friedman (Ed.), *On the other hand: New perspectives on American Sign Language* (pp. 215–236). New York: Academic Press.

Baker, C. (In press). Eye blinking behavior in a visually monitored language. In P. Siple (Ed.), *Understanding language through sign language research.* New York: Academic Press.

Baker, E. F. (1967). *Man in a trap.* New York: Macmillan.

Baldwin, A. L. (1955). *Behavior and development in childhood.* New York: Dryden Press.

Ballentine, R. (1984). Personal communication, December 7.

Bampton, B., Jones, J., & Mancini, J. (1981). Initial mothering patterns of low-income black primiparas. *Journal of Gynecological Nursing, 10,* 174–178.

Bandler, R., & Grinder, J. (1979). *Frogs into princes.* Moab, UT: Real People Press.

Barber, T. X. (1969). Hypnosis: A scientific approach. New York: Van Nostrand.

Barlow, J. D. (1969). Pupillary size as an index of preference in political candidates. Perceptual and Motor Skills, 28, 587–590.

Barrett-Lennard, G. T. (1959). Dimensions of the client's experience of his therapist associated with personality change. (Doctoral dissertation, University of Chicago.) American Doctoral Dissertations, 1959, p. 146.

Barron, E. (1968). The dream of art and poetry. Psychology Today, 2, 18–23, 66.

Bartak, L., & Rutter, M. (1974). Educational treatment of autistic children. In M. Rutter (Ed.), Infantile autism: Concepts, characteristics, treatment. Edinburgh & London: Churchill Livingstone.

Bartlett, E. S., Faw, T. T., & Liebert, R. M. (1967). The effects of suggestions of alertness and hypnosis on pupilarly response: Report on a single subject. International Journal of Clinical and Experimental Hypnosis, pp. 189–192.

Bartlett, J. (1968). Familiar quotations. Boston, MA: Little, Brown.

Bataile, G. (1973). Madame Edwarda: Le mort histoire de l'oeil. Paris: General D'Editions.

Bate, B. R. (1972). Effects of various social reinforcers on interviewee contact. (Doctoral dissertation, Case Western Reserve University, 1972.) Dissertation Abstracts International, 32 (12-B), 7286. (University Microfilms No. 72-18670, 1–190.)

Beattie, G. W. (1978). Sequential temporal patterns of speech and gaze dialogue. Semiotica, 23, 29–52.

Beattie, G. W. (1979). Planning units in spontaneous speech: Some evidence from hesitations in speech and speaker gaze direction in conversation. Linguistics, 17, 61–78.

Beier, E. G., & Sternberg, D. P. (1977). Marital communication. Journal of Communication, 27, 92–97.

Benjamin, J. D. (1963). Further comments on some developmental aspects of anxiety. In H. Gaskill (Ed.), Counterpoint. New York: International Universities Press.

Berger, J. (1980). About looking. New York: Pantheon Books.

Bergman, E. (1976). Art as a medium of communication. Proceedings of the Second Gallaudet Symposium on Research in Deafness (pp. 124–126). Washington, DC: Gallaudet Press.

Berryman, J. (1971). "#385." In A. Poulin (Ed.), Contemporary American poetry (p. 26). Boston, MA: Houghton Mifflin.

Betteridge, H. T. (1962). The new Cassell's German dictionary (p. 85). New York: Funk & Wagnalls.

Birdwhistell, R. L. (1952). Field methods and techniques. Human Organization, 11, 37–38.

Birdwhistell, R. L. (1964). Communication as a multichannel system. In D. L. Sills (Ed.), International encyclopedia of the social sciences (Vol. 3, pp. 24–28). New York: Macmillan and The Free Press.

Birdwhistell, R. L. (1970). Kinesics and context: Essays on body motion communication. Philadelphia: University of Pittsburgh Press.

Biren, J. (JEB). (1979). Eye to eye: Portraits of lesbians. Washington, DC: Glad Hag Books.

Birge, H. L. (1945). Ocular war neuroses. Archives of Opthalmology, 33, 440–468.

Bishop, M. (1975). A funeral for the eyes of fire. New York: Ballantine.

Blair, L. (1976). Rhythms of vision: The changing patterns of beliefs. New York: Schocken.

Blake, P., & Moss, T. (1967). The development of socialization skills in an electively mute child. Behavior Research and Therapy, 5, 349–356.

Blank, H. R. (1963). Dream analysis in the treatment of the blind. In E. T. Adelson (Ed.), *Dreams in contemporary psychoanalysis* (p. 188). New York: Society of Medical Psychoanalysts.

Blank, L., Gottsegen, G. B., & Gottsegen, M. G. (1971). *Confrontation: Encounters in self and interpersonal awareness.* New York: Macmillan.

Blasband, R. A. (1967). The significance of the eye block in psychiatric Orgone Therapy. *Journal of Orgonomy, 1,* 156–163.

Blondis, M. N., & Jackson, B. E. (1977). *Nonverbal communication with patients back to the human touch.* New York: Wiley.

Bonny, H. L. (1978a). *Facilitating Guided Imagery and Music sessions.* GIM Monograph #1. Baltimore, MD: ICM Books.

Bonny, H. L. (1978b). *The role of taped music programs in the GIM process.* GIM Monograph #2. Baltimore, MD: ICM Books.

Bonny, H. L., & Savary, L. M. (1973). *Music and your mind: Listening with a new consciousness.* New York: Harper & Row.

Bowlby, J. (1958). The nature of the child's tie to its mother. *International Journal of Psychoanalysis, 39,* 350–373.

Bradley, J. P., Daniels, L. F., & Jones, T. C. (1969). *The international dictionary of thoughts* (pp. 272–273). Chicago, IL: J. G. Fergusen.

Brady, J. P., & Rosner, B. S. (1966). Rapid eye movement in hypnotically induced dreams. *Journal of Nervous and Mental Disease, 143*(1), 28–35.

Braginsky, B. M., & Braginsky, D. D. (1974). *Mainstream psychology, a critique.* New York: Holt, Rinehart & Winston.

Breed, G. R. (1969). *Nonverbal communication and interpersonal attraction in dyads.* (Doctoral dissertation, University of Florida.) *Dissertation Abstracts International,* 31/03A, 1369. (University Microfilms No. 70-14857.)

Breed, G. R. (1972). The effect of intimacy: Reciprocity or retreat? *British Journal of Social and Clinical Psychology, 11,* 135–142.

Brenman, M., & Gill, M. M. (1971). *Hypnotherapy: A survey of the literature.* New York: International Universities Press.

Brill, P. M. (1983). Preference for counselor eye contact from a male counselor by college-age male clients' interpersonal orientations on FIRO-B. (Doctoral dissertation, University of Maine.) *Dissertation Abstracts International, 47,* 2328-B. (University Microfilms No. 82-26726.)

Brinnin, J. M. (Ed.). (1960). *Emily Dickinson.* New York: Dell Publishing.

Broverman, I. K., Broverman, D. M., Clarkson, F. E., & Vogel, S. R. (1970). Sex role stereotypes and clinical mental health. *Journal of Consulting and Clinical Psychology, 34,* 1–7.

Brown, D., & Parks, J. C. (1972). Interpreting non-verbal behavior, a key to more effective counseling: A review of literature. *Rehabilitation Counseling Bulletin, 15,* 176–184.

Brown, H. Rap. (1969). *Die, nigger, die!* New York: Dial Press.

Brown, N. O. (Trans.). (1953). *Hesiod's theogony.* New York: Liberal Arts Press.

Brown, R., & Herrnstein, R. J. (1975). *Psychology.* Boston, MA: Little, Brown.

Brown, R. M. (1974). The good fairy. *Quest: A Feminist Quarterly, 1,* 58–64.

Brown, S. R. (1972). Eye contact as an indicator of infant social development. (Doctoral dissertation, University of Southern California.) *Dissertation Abstracts International,* 33/02B. (University Microfilms No. 72-21654, p. 909.)

Bruner, J. S. (1967). *Eye, hand, and mind.* (Unpublished manuscript, Harvard University.)

Bruner, J. S. (1968). *Process of cognitive growth in infancy.* Heinz Werner Lectures, Clark University, Worcester, February 8–9.

Bruton, P. (1975). *The quest of the overself*. New York: Samuel Weiser.

Buchanan, D. R., Goldman, M., & Juhnke, R. (1977). Eye contact, sex, and the violation of personal space. *The Journal of Social Psychology, 103,* 19–25.

Buchheimer, A. (1963). The development of ideas about empathy. *Journal of Counseling Psychology, 10,* 61–70.

Buck, R. W., Savin, V. J., Miller, R. E., & Caul, W. F. (1972). Communication of affect through facial expressions in humans. *Journal of Personality and Social Psychology, 3,* 362–371.

Budge, E. A. W. (1969). *The gods of the Egyptians* (Vol. 1). New York: Dover Publications.

Buffington, P. W. (1984, February). The psychology of the eyes. *Sky,* 92–96.

Buhler, C., & Hetzer, H. (1928). Veherste Verstandnis fur Ausdruck im ersten Lebensjahr. *Zeitschrift fur Psychologie, 107,* 50–61.

Bulfinch, T. (1968). *Bulfinch's mythology.* Garden City, NY: Doubleday.

Bullock, A. (1960). *Hitler: A study in tyranny.* New York: Harper.

Burroughs, W., Schultz, W., & Autrey, S. (1973). Quality of argument, leadership votes, and eye contact in three person leaderless groups. *Journal of Social Psychology, 90,* 89–93.

Burton, M. (1957, October 27). Birds fear the evil eye. *Illustrated London News,* 712.

"Busting the Bhagwan." (1985, Nov. 11). *Newsweek,* pp. 26, 31–32.

Buzby, D. E. (1924). The interpretation of facial expression. *American Journal of Psychology, 35,* 602–604.

Callan, H. M. W., Chance, M. R. A., & Pitcairn, T. K. (1973). Attention and advertence in human groups. *Social Science Information, 12,* 27–41.

Campbell, J. (1968). *The masks of God: Creative mythology.* New York: Penguin Books.

Cantin, J. (1976). The cat in Egypt. In J. Fireman (Ed.), *The cat catalog* (pp. 3–9). New York: Workman.

Caproni, V., Levine, D., O'Neal, E., McDonald, P., & Garwood, G. (1977). Seating position, instructor's eye contact availability, and the student participation in a small seminar. *Journal of Social Psychology, 103,* 315–316.

Carpenter, G. C., & Stechler, G. (1967). Selective attention to the mother's face from week 1 through week 8. *Proceedings of the 75th Annual Convention of the American Psychological Association,* 153–154.

Cary, M. (1978). The role of gaze in the initiation of conversation. *Social Psychology, 3,* 269–271.

Cary, M. (1979). Gaze and facial display in pedestrian passing. *Semiotica, 28,* 323–326.

Cegala, D. J., Alexander, A. F., & Sokuvitz, S. (1979). An investigation of eye gaze and its relation to selected verbal behavior. *Human Communication Research, 5,* 99–108.

Chance, M. R. A. (1967). Attention structure as the basis of primate rank order. *Man, 2,* 503–518.

Chapman, A. J. (1975). Eye contact, physical proximity and laughter: A re-examination of the equilibrium model of social intimacy. *Social Behavior and Personality, 3,* 143–155.

Chapple, E. D. (1940). "Personality" differences as described by invariant properties of individuals in interaction. *Proceedings of the National Academy of Sciences, 26,* 10–16.

Cherulnik, P. O., Neely, W. T., Flanagan, M., & Zachau, M. (1979). Social skill and visual interaction. *Journal of Social Psychology, 104,* 263–270.

Chevalier-Skolnikoff, S. (1973). Facial expression of emotion in non-human primates. In P. Eckman (Ed.), *Darwin and facial expression.* New York: Academic Press.

Christensen, P. (1978). *A tempered life.* (Unpublished manuscript.)

Christie, R., & Geis, F. (1968). Consequences of taking Machiavelli seriously. In E. F. Borgatta & W. W. Lambert (Eds.), *Handbook of Personality Theory and Research* (pp. 959–973). Chicago: Rand McNally.

Cirlot, J. (1962). *A dictionary of symbols*. New York: Philosophical Library.

Clancy, J. (1963). Eye complaints as symptoms of psychiatric disorder. *American Journal of Opthalmology, 55*, 767–773.

Clark, E. (1977). Eyes, etc.: A memoir. New York: Random House.

Cline, M. (1967). The perception of where a person is looking. *American Journal of Psychology, 80*, 41–50.

Cohen, D. (1979, October). Communication: The avid gaze of strangers. *Psychology Today, 40*, 115.

Cohen, L. J., & Campos, J. J. Father, mother, and strangers as elicitors of attachment behaviors in infancy. *Developmental Psychology, 10*, 146–154.

Cohen, M. (1949). Preliminary report on ocular findings in 323 schizophrenic patients. *Psychiatric Quarterly, 23*, 667–671.

Collett, P. (1971). On training Englishmen in the non-verbal behavior of Arabs: An experiment in intercultural communication. *International Journal of Psychology, 6*, 209–215.

Colligan, R. C., & Bellamy, C. M. (1968). Effects of a two year treatment program for a young autistic child. *Psychotherapy: Theory, Research and Practice, 5*, 214–219.

Condon, J. C., & Yousef, F. (1975). *An introduction to intercultural communication*. New York: Bobbs-Merrill.

Condon, W., & Ogston, W. D. (1966). Sound film analysis of normal and pathological behavior patterns. *Journal of Nervous Mental Disorders, 143*, 338–347.

Condon, W. G., & Ogston, W. D. (1971). Speech and body motion synchrony of the speaker-hearer. In D. L. Horton & J. J. Jenkins (Eds.), *The Perception of Language*. Columbus, OH: Charles E. Merrill.

Condon, W. S., Ogston, W. D., & Pacoe, L. V. (1969). Three faces of Eve revisited: A study of transient microstrabismus. *Journal of Abnormal Psychology, 74*, 618–620.

Cook, M. (1970). Experiments on orientation and proxemics. *Human Relations, 23*, 61–76.

Cook, M., & Smith, J. M. C. (1975). The role of gaze in impression formation. *British Journal of Social and Clinical Psychology, 14*, 19–25.

Cott, H. B. (1957). *Adaptive coloration in animals*. New York: Methuen.

Coutts, L. M., & Schneider, F. W. (1975). Visual behavior in an unfocused interaction as a function of sex and distance. *Journal of Experimental Social Psychology, 11*, 64–77.

Cranach, M. von. (1971). The role of orienting behavior in human interaction. In A. H. Esser (Ed.), *Behavior and environment: The use of space by animals and men.* (pp. 217–237). New York: Plenum Press.

Cranach, M. von, & Ellgring, J. H. (1973). The perception of looking behaviour. In M. von Cranach & I. Vine (Eds.), *Social communication and movement*. London: Academic Press.

Cranach, M. von, & Vine, I. (1973). Introduction. In M. von Cranach & I. Vine (Eds.), *Social communication and movement*. London: Academic Press.

Crosby, F., Jose, P., & Wong-McCarthy. (1981). Gender, androgeny, and conversational assertiveness. In C. Mayo & N. M. Henley (Eds.), *Gender and nonverbal behavior*. New York: Springer-Verlag.

Dabbs, J. M., Jr. (1969). Similarity of gestures and interpersonal influence. *Summary in Proceedings of the 77th Annual Convention of the American Psychological Association* (Vol. 4, pp. 337–338). Washington, DC: American Psychological Association.

Dabbs, J. M., Evans, M. S., Hopper, C. H., & Purvis, J. K. (1980). Self-monitors in conversation: What do they monitor? *Journal of Personality and Social Psychology,* 39, 278–284.

Dabbs, J. M., Johns, C. J., & Powell, P. M. (1976). *Less eye contact when closer? Depends upon your partner's sex.* Unpublished manuscript submitted to American Psychological Association, January 30, 1976.

Daly, M. (1978). *Gyn/ecology: The metaethics of radical feminism.* Boston, MA: Beacon Press.

Dames, M. (1976). *The Silbury treasure.* London: Thames & Hudson.

Daniell, R. J., & Lewis, P. (1972). Stability of eye contact and physical distance across a series of structured interviews. *Journal of Consulting and Clinical Psychology,* 39, 172.

Darwin, C. (1979). *The expression of the emotions in man and animals.* London: Julian Friedman.

Davis, F. (1971). *Inside intuition: What we know about nonverbal communication.* New York: McGraw-Hill.

Davis, M. (1970). Movement characteristics of hospitalized psychiatric patients. In *Proceedings of the fifth annual conference of the American Dance Therapy Association,* 25–45.

Davis, M. (1979a). Laban analysis of nonverbal communication. In Shirley Weitz (Ed.), *Nonverbal communication: Readings and commentary.* New York: Oxford University Press.

Davis, M. (1979b). The jury work of Rosalyn Lindner. *The Kinesis Report,* 2, 54.

Davy, J. (1981, August 30). What if Darwin were wrong? *The Washington Post,* C1, 3.

Day, M. E. (1967). An eye-movement indicator of individual differences, the physiological organization of attentional processes and anxiety. *The Journal of Psychology,* 65, 51–62.

de Purucker, G. (1974). *Fountain—Source of occultism.* Pasadena, CA: Theosophical University Press.

de Riencourt, A. (1981). *The eye of Shiva.* New York: William Morrow.

de Vries, A. (1974). *Dictionary of symbols and imagery.* Amsterdam: North-Holland.

Dew, R. A. (1970). The Biopathic diathesis (VII): Headache. *Journal of Orgonomy,* 8, 143–154.

Dickens, C. (1977). *Bleak house* (An authoritative and annotated text.) (G. Ford & S. Monad, Eds.). New York: W. W. Norton. (Original work published 1852).

Dickinson, E. (1970). *The complete poems of Emily Dickinson* (Thomas H. Johnson, Ed.). London: Faber & Faber.

Diebold, A. R. (1968). Biology of the comparative psychology of communicative behavior. In T. A. Sebeok (Ed.), *Animal communication—Techniques of study and results of research* (pp. 525–571). Bloomington, IN: Indiana University Press.

Diefendorf, A. R., & Dodge, R. (1908). Experimental study of ocular reactions of the insane from photographic records. *Brain,* 31, 451–489.

Dittman, A. (1978). The role of body movement in communication. In A. Seigman & D. Feldstein (Eds.), *Nonverbal behavior and communication.* Hillsdale, NJ: Lawrence Erlbaum Associates.

Dodson, E. D., & Dodson, P. (1976). *Evolution: Process and product.* New York: Van Nostrand.

Does he look at books? (1977, November). *Mother Jones,* 9.

Dollard, J., & Miller, N. E. (1950). *Personality and psychotherapy: An analysis in terms of learning, thinking and culture.* New York: McGraw-Hill.

Donaldson, C. (Ed.). (1974). *Understanding human behavior* (Vol. 2). New York: B.C.P. Publishing Co.

Donovan, W. L., & Leavitt, L. A. (1980). Physiologic correlates of direct and averted gaze. *Biological Psychology, 10*, 189–199.

Dovidio, J. F., & Ellyson, S. L. (1982). Decoding visual dominance: Attributions of power based on relative percentages of looking while speaking and looking while listening. *Social Psychology Quarterly, 45*, 106–113.

Drag, R. (1968). *Self-disclosure as a function of group size and experimenter behavior.* (Doctoral dissertation, University of Florida.)

Dubos, R. (1968). *So human an animal.* New York: Charles Scribner.

Duncan, S. (1969). Nonverbal communication. *Psychology Bulletin, 72*, 118–137.

Dunlap, K. (1927). The role of eye-muscles in the expression of emotions. *Genetic Psychology Monographs, 2*, 199–233.

Eder, M. D. (1913). Augentraume. *Internationale Zeitschrift für Psychoanalyse, 1*, 157–158.

Efran, J. S. (1968). Looking for approval: Effects on visual behavior of approbation from persons differing in importance. *Journal of Personality and Social Psychology, 10*, 21–25.

Efran, J. S., & Broughton, A. (1966). Effect of expectancies for social approval on visual behavior. *Journal of Personality and Social Psychology, 4*, 103–107.

Eibl-Eibesfeldt, I. (1972). Similarities and differences between cultures in expressive movements. In R. A. Hinde (Ed.), *Nonverbal communication.* Cambridge: Cambridge University Press.

Eibl-Eibesfeldt, I. (1973). Expressive behavior of the deaf and blind born. In M. von Cranach & I. Vine (Eds.), *Social communication and movement* (pp. 163–194). New York: Academic Press.

Eisler, R. M., Hersen, M., & Agras, S. W. (1973). A method for the controlled observation of nonverbal interpersonal behavior (videotape). *Behavior Therapy, 4*, 420–425.

Ekman, P. (1965). Communication through nonverbal behavior: A source of information about an interpersonal relationship. In S. S. Thompkins & C. E. Izard (Eds.), *Affect, cognition, and personality.* New York: Springer Publishing.

Ekman, P. (1972). Universal and cultural differences in facial expression of emotions. In J. Cole (Ed.), *Nebraska Symposium on Motivation* (Vol. 19). Lincoln, NE: University of Nebraska Press.

Ekman, P. (Ed.). (1973) *Darwin and facial expression: A century of research in review.* New York: Academic Press.

Ekman, P. (1977). Personal communication, August 20.

Ekman, P., & Friesen, W. V. (1967). Head and body cues in the judgment of emotion: A reformulation. *Perceptual and Motor Skills, 24*, 711–724.

Ekman, P., & Friesen, W. V. (1969). Nonverbal leakage and clues to deception. *Psychiatry, 32*, 88–106.

Ekman, P., & Friesen, W. V. (1975). *Unmasking the face.* Englewood Cliffs, NJ: Prentice-Hall.

Ekman, P., & Friesen, W. V. (1976). Measuring facial movement. *Environmental Psychology and Nonverbal Behavior, 1*, 56–75.

Ekman, P., Sorenson, E. R., & Friesen, W. V. (1969). Pancultural elements in facial displays of emotions. *Science, 164*, 86–88.

Eliot, T. S. (1963). *Collected poems 1909–1962.* New York: Harcourt, Brace, & World.

Ellgring, J. H. (1970). Judgment of glances directed at different points in the face. *Zeitschrift für Experimentelle und Angewandte Psychologie, 17*, 600–607.

Ellgring, J. H., & von Cranach, M. (1972). Process of learning in the recognition of eye-signals. *European Journal of Social Psychology, 2*, 33–43.

Ellison, H. (1975). *Deathbed stories.* New York: Harper & Row.

Ellsworth, P. G. (1975). Direct gaze as a social stimulus: The example of aggression. In P. Pliner, L. Krames, & T. Alloway (Eds.), *Nonverbal communication of aggression* (pp. 53–76). New York: Plenum Press.

Ellsworth, P. C., & Carlsmith, J. M. (1968). Effects of eye contact and verbal contact on affective response to a dyadic interaction. *Journal of Personality and Social Psychology, 10*, 15–20.

Ellsworth, P. C., & Carlsmith, J. M. (1973). Eye contact and gaze aversion in an aggressive encounter. *Journal of Personality and Social Psychology, 28*, 280–292.

Ellsworth, P. C., Carlsmith, J. M., & Henson, A. (1972). The stare as a stimulus to flight in human subjects: A series of field experiments. *Journal of Personality and Social Psychology, 21*, 302–311.

Ellsworth, P. C., & Langer, E. J. (1976). Staring and approach: An interpretation of the stare as a nonspecific activator. *Journal of Personality and Social Psychology, 33*, 117–122.

Ellsworth, P. C., & Ludwig, L. M. (1972). Visual behavior in social interaction. *Journal of Communication, 22*, 375–403.

Ellsworth, P. C., & Ross, L. D. (1975). Intimacy in response to direct gaze. *Journal of Experimental Social Psychology, 11*, 592–613.

Ellyson, S. L. (1975). Visual behavior exhibited by males differing as to interpersonal control orientation in one- and two-way communication systems. (Doctoral dissertation, University of Delaware, 1974.) Cited in R. V. Exline, S. L. Ellyson, and B. Long. Visual behavior as an aspect of power role relationships. In P. Pliner, L. Krames, & T. Alloway (Eds.), *Nonverbal communication of aggression.* New York: Plenum.

Ellyson, S. L., Dovidio, J. F., & Fehr, B. J. (1981). Visual behavior and dominance in women and men. In C. Mayo & N. Henley (Eds.), *Gender and nonverbal behavior.* New York: Springer-Verlag.

Elman, D. (1973). Eye contact, interest, and arousal. (Doctoral dissertation, Columbia University, 1973.) *Dissertation Abstracts International, 34*, 2765A. (University Microfilms No. 73-26604.)

Elman, D., Schulte, D. C., & Bukoff, A. (1977). Effects of facial expression and stare duration on walking speed: Two-fold experiments. *Environmental Psychology and Nonverbal Behavior, 2*, 93–99.

Epting, F. R., Suchman, D. I., & Barker, E. N. (1969). Some aspects of revealingness and disclosure: A review. (Unpublished manuscript, University of Florida.)

Erickson, F. (1976, May 11). Talking down and giving reasons: Hyper-explanation and listening behavior in inter-racial interviews. (Paper delivered at the International Conference on Nonverbal Behavior, Ontario Institute for Studies in Education, Toronto, Canada.)

Escher, M. C., & Locher, J. L. (Eds.). (1971). *The World of M. C. Escher.* New York: Harry N. Abrams.

Etkin, W. (1967). *Social behavior from fish to man.* Chicago: University of Chicago Press.

Etzel, B. C., & Gerwitz, J. L. (1967). Experimental modifications of caretaker-maintained high-rate operant crying in a 6- and 20-week-old infant (*Infans Tyrannotenus*): Extinction of crying with reinforcement of eye-contact and smiling. *Journal of Experimental Child Psychology, 5*, 303–317.

Evans, H. C. (1925). The evil eye. *Psyche, 22*, 101–106.

Exline, R. V. (1962). Need affiliation and initial communication behavior in task-oriented groups characterized by low interpersonal visibility. *Psychological Reports, 10*, 78–89.

Exline, R. V. (1963). Explorations in the process of person perception: Visual interaction in relation to competition, sex, and need for affiliation. *Journal of Personality,* 31, 1–20.

Exline, R. V. (1971). Visual interaction: The glances of power and preference. In J. Cole (Ed.), *Nebraska symposium on motivation* (Vol. 19, pp. 163–206). Lincoln: University of Nebraska Press.

Exline, R. V., & Eldridge, C. (1967). *Effects of two patterns of a speaker's visual behavior upon the perception of the authenticity of his verbal message.* Paper presented at the annual convention of the Eastern Psychological Association, Boston.

Exline, R. V., Ellyson, E. L., & Long, B. (1975). Visual behavior as an aspect of power role relationships. In P. Pliner et al. (Eds.), *Nonverbal communication of aggression* (pp. 21–52). New York: Plenum.

Exline, R. V., & Fehr, B. (1978). Applications of semiosis to the study of visual interaction. In A. Seigman & D. Feldstein (Eds.), *Nonverbal behavior and communication* (pp. 117–158). Hillsdale, NJ: Lawrence Erlbaum Associates.

Exline, R. V., & Fehr, B. J. (1982). The assessment of gaze and mutual gaze. In K. R. Sherer and P. Ekman (Eds.), *Handbook of methods of nonverbal behavior.* London: Cambridge University Press.

Exline, R. V., Gottheil, I., Paredes, A., & Winklemeier, D. (1968). Gaze direction as a factor in the accurate judgment of nonverbal expressions of affect. *Proceedings of the 76th Annual Convention of the American Psychological Association,* 3, 415–416.

Exline, R. V., Gray, D., & Schuette, D. (1965). Visual behavior in a dyad as affected by interview content and sex of respondent. *Journal of Personality and Social Psychology,* 1, 201–209.

Exline, R. V., Jones, P., & Maciorowski, K. (1977). *Race, affiliation—conflict theory and mutual visual attention during conversation.* (Paper presented at the annual convention of the American Psychological Association, San Francisco.)

Exline, R. V., Lanzetta, J. T., Pruitt, D. G., & Ziller, R. C. (1969). Experimental studies of social interaction—perception variables and communication efficiency. (Center for Research on Social Behavior, University of Delaware, Newark, DE.)

Exline, R. V., & Messick, D. (1967). The effects of dependency and social reinforcement upon visual behavior during an interview. *British Journal of Social and Clinical Psychology,* 6, 256–266.

Exline, R. V., Thibaut, J., Brannon, C., & Gumpert, P. (1961). Visual interaction in relation to Machiavellianism and an unethical act. *American Psychologist,* 16, 396.

Exline, R. V., Thibaut, J., Hickey, C. B., & Gumpert, P. (1970). Visual interaction in relation to Machiavellianism and an unethical act. In R. Christie & F. Geiss, *Studies in Machiavellianism* (pp. 53–75). New York: Academic Press.

Exline, R. V., & Winters, L. C. (1965a). Affective relations and mutual glances in dyads. In S. S. Thompkins & C. E. Izards (Eds.), *Affect, cognition, and personality* (pp. 319–350). New York: Springer Publishing.

Exline, R. V., & Winters, L. C. (1965b). The effects of cognitive difficulty and cognitive style upon eye to eye contact in interviews. (Paper presented at the annual convention of the Eastern Psychological Association, Atlantic City, NJ.)

Exline, R. V., & Yellin, A. (1971). Eye contact as a sign between man and monkey. *Proceedings of the XIXth International Congress of Psychology.* London.

Eyes. (1970). *Voices:* special issue, 105.

Eysenck, H. J. (1970). Causal theories of personality structure. In *The structure of human personality* (pp. 425–463). London: Methuen.

Falzett, W. C., Jr. (1981). Matched versus unmatched primary representational systems and their relationship to perceived trustworthiness in a counseling analogue. *Journal of Counseling Psychology, 28*, 305–308.

Fantz, R. L. (1958). Pattern vision in young infants. *Psychological Record, 8*, 43–47.

Fantz, R. L. (1961). The origin of form perception. *Scientific American, 5*, 66–72.

Farren, D., Hirschbiel, P., & Jay, S. (1980). Toward interactive synchrony: The gaze patterns of mothers and children in three age groups. *International Journal of Behavioral Development, 3*, 215–224.

Fast, J. (1970). *Body Language.* New York: M. Evans.

Fast, J. (1977). *The body language of sex, power, and aggression.* New York: M. Evans.

Fast, J. (1978). Excuse me, but your eyes are talking. *Family Health, 10*, 22–25.

Fehr, B. J., & Exline, R. V. (1978). *Visual interaction in same-and interracial dyads.* (Paper presented at the Eastern Psychological Association Meetings, Washington, DC.)

Fehr, B. J., & Exline, R. V. (in press). *Social visual interaction: A conceptual and literature review.* Hillsdale, NJ: Lawrence Erlbaum Associates.

Feldman, R. S., & Orchowsky, S. (1979). Race and performance of student as determinants of teacher nonverbal behavior. *Contemporary Educational Psychology, 4*, 324–333.

Firestone, S. (1970). *The dialectic of sex.* New York: Morrow.

Fitzgerald, F. S. (1925). *The great Gatsby.* New York: Grosset & Dunlap.

Forrest, E. B. (1969, January). A dynamic model of vision. *American Journal of Optometry*, 37–44.

Fowler, A. (1966). *Fitzgerald's the great Gatsby.* New York: Barrister Publishing.

Fraiberg, S. (1974). Blind infants and their mothers: An examination of the sign system. In M. Lewis and L. A. Rosenblum (Eds.), *The origins of behavior*, Vol. 1: *The effect of the infant on its caregiver.* New York: Wiley.

Frame, C., Matson, J. L., Sonis, W. A., Fialkor, M. J., & Kazdin, A. E. (1982). Behavioral treatment of depression in a prepubertal child. *Journal of Behavior Therapy and Experimental Psychiatry, 13*, 239–43.

Freedman, D. G. (1965). Hereditary control of early social behavior. In I. M. Foss (Ed.), *Determinants of infant behavior* (pp. 213–299). New York: Wiley.

Freedman, D. G. (1979). *Human sociobiology: A holistic approach.* New York: Free Press.

Freedman, J. L. (1975). *Crowding and behavior.* San Francisco: W. H. Freedman.

Fretz, B. R., Corn, R., Tuemmler, J. M., & Bellett, W. (1979). Counselor nonverbal behaviors and client evolutions. *Journal of Counseling Psychology, 26*, 304–311.

Freud, S. (1955). *The interpretation of dreams.* London: Hogarth Press.

Friedman, N. (1965). *The psychological experiment as a social interaction.* (Doctoral dissertation, Harvard University.)

Friessen, W. (1972). Cultural differences in facial expressions in a social situation: An experimental test of the concept of display rules. *Dissertation Abstracts International, 33/08*, 3976. (University Microfilms No. DCJ 73-03651.)

Frieze, I. H. (1978). Being feminine or masculine—nonverbally. In I. H. Frieze, J. E. Parsons, P. B. Johnson, N. Ruble, & G. L. Zellman (Eds.), *Women and sex roles: A social psychological perspective.* New York: W. W. Norton.

Frieze, I. H., & Ramsey, S. J. (1976). Nonverbal maintenance of traditional sex roles. *Journal of Social Issues, 32*, 133–141.

Fromme, D. K., & Beam, D. C. (1974). Dominance and sex differences in non-verbal responses to differential eye contact. *Journal of Research in Personality, 8*, 76–87.

Fromme, D. K., & Schmidt, C. (1972). Affective role enactment and expressive behavior. *Journal of Personality and Social Psychology, 24*, 413–19.

Frost, C. (1971). "Acquainted with the night." In C. Sanders, R. R. Rice, & W. J. Cantillon (Eds.), *Synthesis: Responses to Literature*. New York: Alfred A. Knopf.

Fugita, S. S. (1974). Effects of anxiety and approval on visual interaction. *Journal of Personality and Social Psychology, 29*, 586–592.

Fugita, S. S., Agle, T. A., Newman, I., & Walfish, N. (1977). Attractiveness, self-concept, and a methodological note about gaze behavior. *Personality and Social Psychology Bulletin, 3*, 240–243.

Fujita, B. N., Harper, R. G., & Wiens, A. N. (1980). Encoding-decoding of nonverbal emotional messages: Sex differences in spontaneous and enacted expressions. *Journal of Nonverbal Behavior, 4*, 131–145.

Gage, M. J. (1980). *Woman, church and state*. Watertown, MA: Persephone Press.

Galassi, J. P., et al. (1976). Behavioral performance in the validation of an assertiveness scale, *Behavior Therapy, 7*, 447–452.

Gale, A., Lucas, B., Nissin, R., & Harpham, B. (1972). Some EEG correlates of face-to-face contact. *British Journal of Social and Clinical Psychology, 11*, 326–332.

Gale, A., Spratt, G., Chapman, A. J., & Smallbone, A. (1975). EEG correlates of eye contact and interpersonal distance. *Biological Psychology, 3*, 237–245.

Gardner, H. (1980). *Art through the ages*. New York: Harcourt, Brace, Jovanovich.

Garrison, V., & Arensberg, C. (1976). The evil eye: Envy or risk of seizure? Paranoia or patronal dependency? In Clarence Maloney (Ed.), *The evil eye*. New York: Columbia University Press.

Gaskell, G. A. (1960). *Dictionary of all scriptures and myths*. New York: Julian Press.

Gaskin, S. (1972). *Caravan*. New York & Berkeley, CA: Random House & The Bookworks.

Gaskin, S. (1984). Personal communication, May.

Gesell, A. L., Ilg, F. L., & Ames, L. B. (1956). *Youth: The years from ten to sixteen*. New York: Harper & Row.

Gesell, A. L., Ilg, F. L., & Bullis, G. E. (1949). *Vision*. New York: Harper & Row.

Gewirtz, H., & Gewirtz, J. L. (1969). Caretaking settings, background events and behavior differences in four Israeli child-rearing environments: Some preliminary trends. In B. M. Foss (Ed.), *Determinants of infant behavior* (Vol. 4, pp. 229–252). London: Methuen.

Gibson, J. J., & Pick, A. D. (1963). Perception of another person's looking behavior. *American Journal of Psychology, 76*, 386–394.

Giessler, C. M. (1913). Der Blick des Menschen als Ausdruck seins Seelenlebens. *Zeitschrift für Psychologie, 65*, 181–211.

Goffman, E. (1959). *The presentation of self in everyday life*. New York: Anchor Books.

Goffman, E. (1963a). *Behavior in public places*. Glencoe, IL: Free Press.

Goffman, E. (1963b). *Stigma: Notes on the management of spoiled identity*. Englewood Cliffs, NJ: Prentice-Hall.

Goldberg, G. N., Kiesler, C. A., & Collins, B. E. (1969). Visual behavior and face-to-face distance during interaction. *Sociometry, 32*, 43–53.

Goldschmidt, W. (1975). Absent eyes and idle hands: Socialization for low affect among the Sebei. *Ethos*, Summer, 157–163.

Goldstein, M. A., Kilroy, M. C., & Van de Voort, D. (1976). Gaze as a function of conversation and degree of love. *Journal of Psychology, 92*, 227–234.

Goodman, R. T. (1976). "My eyes find words." *Poet Lore, 71*, 52.

Gordon, S. (1975). *Two-eyes*. London: Sidgwick & Jackson.

Graves, R. (1948). *The white goddess*. New York: Farrar, Straus, & Giroux.

Gray, S. L. (1971). Eye contact as a function of sex, race, and interpersonal needs. (Doctoral dissertation, Case Western Reserve University.) *Dissertation Abstracts International, 32*, 1842B. (University Microfilms No. 71-22, 805.)

Greenacre, P. (1926). The eye motif in delusion and fantasy. *American Journal of Psychiatry*, 5, 553–579.

Greenbaum, P. (1977). The effects of interpersonal distance, staring and the knitted brow on recipient approach avoidance behaviors in a field setting. (Doctoral dissertation, University of Kansas.)

Greenberg, J. (1976). The role of seating position in a group interaction: A review, with application for group trainers. *Group Organizational Studies*, 1, 310–327.

Greenfield, S. C. (1979). Navajo-anglo nonverbal communication. *Journal of the Arizona Communication & Theatre Association*, Spring, 1–3.

Greenman, G. W. (1963). Visual behavior of newborn infants. In A. J. Solnit & S. Provence (Eds.), *Modern perspectives in child development*. New York: International Universities Press.

Greenson, R. R. (1961). On the silence and sounds of the analytic hour. *Journal of the American Psychoanalytic Association*, 9, 79–84.

Greenson, R. R. (1967). *The technique and practice of psychoanalysis.* New York: International Universities Press.

Gregory, L. (1976). *Visions and beliefs in the west of Ireland.* Toronto: Macmillan.

Griffen, B. Q., & Rogers, R. W. (1977). Reducing interracial aggression. Inhibiting effects of victim's suffering and power to retaliate. *Journal of Psychology*, 95, 151–157.

Griffith, A. V., & Peyman, D. A. R. (1959). Eye-ear emphasis in the DAP as indicating ideas of reference. *Journal of Consulting Psychology*, 23, 560.

Grinder, J., DeLozier, J., & Bandler, R. (1977). *Patterns of the hypnotic techniques of Milton H. Erickson, M.D.* (Vol. 2). Cupertino, CA: Meta Publications.

Grinker, R. R., et al. (1961). *The phenomena of depressions.* New York: Hoeber.

Grosshans, O. R. (1978). A student teacher checklist. *Health Education*, 9, 30–31.

Grumet, G. W. (1983, May). Eye contact: The core of interpersonal relatedness. *Psychiatry*, 46, 172–180.

Guthrie, R. D. (1976). *Body hot spots: The anatomy of human social organs and behavior.* New York: Van Nostrand Reinhold.

Haase, R. F., & Tepper, D. T., Jr. (1972). Nonverbal components of empathic communication. *Journal of Counseling Psychology*, 19, 417–424.

Haeckel, E. (1900). *The riddle of the universe.* New York: Harper & Brothers.

Haggard, E. A., & Issacs, F. S. (1966). Micromomentary facial expressions as indicators of ego mechanisms in psychotherapy. In L. A. Gottschalk & A. H. Auerback (Eds.), *Methods of research in psychotherapy*. New York: Appleton-Century-Crofts.

Hall, E. T. (1959). *The silent language.* Greenwich, CT: Fawcett.

Hall, E. T. (1963). A system for the notation of proxemic behavior. *American Anthropologist*, 65, 1003–1026.

Hall, E. T. (1973). *The silent language.* Garden City, NY: Doubleday.

Hall, E. T., interviewed by K. Friedman. (1979, August). Learning the Arab's silent language. *Psychology Today*, 45–54.

Hall, E. T. (1979). Proxemics. In S. Weitz (Ed.), *Nonverbal communication readings with commentary*. New York: Oxford University Press.

Hall, J. A. (1976). Encoding and decoding of spontaneous and posed facial expressions. *Journal of Personality and Social Psychology*, 34, 966–977.

Hall, J. A. (1978). Gender effects in decoding nonverbal cues. *Psychological Bulletin*, 85, 845–857.

Hall, K. R. (1967). Social interactions of the adult males and adult females of a Patus monkey group. In S. A. Altman (Ed.), *Social communication among primates*. Chicago, IL: University of Chicago Press.

Hamilton, E. (1942). *Mythology: Timeless tales of Gods and heroes.* New York: New American Library.

Hamilton, E. (1969). *Mythology.* New York: Mentor.

Hamm, B. H., & Hartsfield, S. L. (1970). Motivation influencing students in psychiatric nursing. *Nursing Research, 19,* 79–81.

Hammond, D. C., Hepworth, D. H., & Smith, V. G. (1977). *Improving therapeutic communication.* San Francisco: Jossey Bass.

Hanawalt, N. G. (1944). The role of the upper and lower parts of the face as a basis of judging facial expressions: II. In posed expressions and candid camera pictures. *Journal of General Psychology, 31,* 23–36.

Hardee, B. B. (1976). Interpersonal distance, eye contact, and stigmatization: A test of the equilibrium model. (Doctoral dissertation, Johns Hopkins University.) *Dissertation Abstracts International, 37,* B-4. (University Microfilms No. 76-22.)

Harper, R., Weins, A., & Matarazzo, J. (1978). *Nonverbal communication: The state of the art.* New York: Wiley.

Harris, S. E. (1978). Schizophrenic mutual glance patterns. *Psychiatry, 41,* 83–91.

Harrison, R. P. (1973). Nonverbal communication. In I. de Solo Pool, W. Schramm, N. Maccohy, F. Fry, E. Parker, & J. L. Fein (Eds.), *Handbook of communication.* Chicago: Rand McNally.

Harrison, R. P., Cohen, A. A., Crouch, W. W., Genova, B. K. L., & Steinberg, M. (1972). The nonverbal communication literature. *Journal of Communication, 22,* 460–476.

Hart, H. H. (1949). The eye in symbol and symptom. *Psychoanalytic Review, 36,* 1–21.

Hartland, J. (1971). *Medical and dental hypnosis and its clinical applications,* 2nd ed. Baltimore: Williams & Wilkins.

Haviland, J. J., & Malatesta, C. Z. (1981). The development of sex differences in nonverbal signals: Fallacies, facts, and fantasies. In C. Mayo & N. M. Henley (Eds.), *Gender and nonverbal behavior.* New York: Springer-Verlag.

Hawkes, J. (1968). *Dawn of the gods.* New York: Random House.

Heaton, J. M. (1968). *The eye: Phenomenology and psychology of function and disorder.* London: Tavistock.

Heider, F. (1958). *The psychology of interpersonal relations.* New York: Wiley.

Hendin, H. (1975). *The age of sensation.* New York: W. W. Norton.

Henle, M. (Ed.). (1976). *Vision and artifact.* New York: Springer Publishing.

Henley, N. M. (1973). Status and sex: Some touching considerations. *Bulletin of the Psychonomic Society, 2,* 91–93.

Henley, N. M. (1973–74). Power, sex, and nonverbal communication. *Berkeley Journal of Sociology, 18,* 1–26.

Henley, N. M. (1976, April). Nonverbal communication as social control. (Paper presented at Pioneers for Century III.)

Henley, N. M. (1977). *Body politics: Power, sex and nonverbal communication.* Englewood Cliffs, NJ: Prentice-Hall.

Henley, N. M. (1978). Changing the body power structure. *Women, 6,* 34–38.

Henley, N., & Freeman, J. (1975). The sexual politics of interpersonal behavior. In J. Freeman (Ed.), *Women: A feminist perspective.* Palo Alto, CA: Mayfield, 391–401.

Heron, J. (1970). The phenomenology of the social encounter: The gaze. *Philosophy and Phenomenological Research, 2,* 243–264.

Hersen, M., Miller, P. M., and Eisler, R. M. (1973). Interaction between alcoholics and their wives: A descriptive analysis of verbal and nonverbal behavior. *Quarterly Journal of Studies on Alcohol, 34,* 516–520.

Hertz, M. (1961). *Frequency tables for scoring Rorschach responses*, 4th ed. Cleveland, OH: Press of Western Reserve University.

Hess, E. H. (1965). Attitude and pupil size. *Scientific American*, 202, 46–54.

Hess, E. H. (1975). *Tell-tale eye: How your eyes reveal hidden thoughts and emotions*. New York: Van Nostrand Reinhold.

Hesse, H. (1957). *Siddhartha*. (Hilda Rosna, trans.) New York: New Directions.

Hesse, H. (1965). *Demian, the story of Emil Sinclair's youth*. (M. Roloff & M. Lebeck, trans.) New York: Harper & Row.

Hinchliffe, M. K., Lancashire, M., & Roberts, F. J. (1971). A study of eye-contact changes in depressed and recovered psychiatric patients. *British Journal of Psychiatry*, 119, 213–215.

Hingston, R. W. (1933). The meaning of animal color and adornment. London: E. Arnold.

Hittelman, J. H., & Dickes, R. (1979). Sex differences in neonatal eye contact time. *Merrill-Palmer Quarterly*, 25, 171–184.

Hittelman, J. H., Dickes, R., O'Donohue, N., & Bloomfield, R. (1980). Sex differences in neonatal eye contact time: A replication. (Paper presented at the American Psychological Association Meetings, Montreal, Canada.)

Hobson, G. N., Strongman, K. T., Bull, D., & Craig, G. (1973). Anxiety and gaze aversion in dyadic encounters. *British Journal of Social Clinical Psychology*, 12, 122–129.

Hochman, S. (1973). The eyes of flesh. In F. Howe & E. Bass (Eds.), *No more masks! An anthology of poetry by women*. Garden City, NY: Anchor Press.

Hodge, R. L. (1971). Interpersonal classroom communication through eye contact. *Theory into Practice*, 10, 264–267.

Holstein, C. M., Goldstein, J. W., & Bem, D. J. (1971). The importance of expressive behavior, involvement, sex, and need approval in inductive liking. *Journal of Experimental Social Psychology*, 7, 534–544.

Homans, G. (1960). *Social behavior: Its elementary forms*. New York: Wiley.

Honey, P. (1976). *Face to face: Business communication for results*. Englewood Cliffs, NJ: Prentice-Hall.

Horn, J. (Ed.). (1977). The evil eye—A stare of entry. *Psychology Today*, 11, 154–156.

Horn, J. C. (1978, November). Campus prisoners of stereotype. *Psychology Today*, 12, 34.

Hovorka, O. V., & Kronfeld, A. (1908–9). *Vergleichende volksmedizin*. Stuttgart: Strecker & Strecker.

*How to do your face*. (1978). New York: Dell.

How to get tough. (1978, July 10). *Newsweek*, 69.

Howell, T. (1973). Who would have thought. In E. W. Schneider, A. L. Walker, & H. E. Childs (Eds.), *The Range of Literature*, 3rd ed. New York: Van Nostrand.

Howes, D. (1966). A word count of spoken English. *Journal of Verbal Learning and Verbal Behavior*, 5, 572–604.

Huebsch, D. A. (1931). Psychoanalysis and eye disturbances. *Psychoanalytic Review*, 18, 166–179.

Hugo, V. (1971). *La legende des siecles, extraits*, tome I. Paris: Librairie Larousse.

Hulme, F. E. (1969). *The history, principles, and practice of symbolism in Christian art*. Detroit: Gale Research Co.

Hutt, C., & Ounsted, C. (1966). The biological significance of gaze aversion, with particular reference to the syndrome of infantile autism. *Behavioral Science*, 11, 346–356.

Hutt, C., & Vaizey, J. (1966). Differential effects of group density on social behavior. *Nature, 209*, 1371–1732.

Hutt, S. J., Hutt, C., & Ounsted, C. (1965). A behavioral and electro-encephalographic study of autistic children. *Journal of Psychiatric Research, 3*, 181–197.

Ickes, W. (1981). Sex role influences in dyadic interaction: A theoretical model. In C. Mayo & N. M. Henley (Eds.), *Gender and nonverbal behavior*. New York: Springer-Verlag.

Ickes, W., & Barnes, R. D. (1978). Boys and girls together—and alienated! On enacting stereotyped sex roles in mixed-sex dyads. *Journal of Personality and Social Psychology, 36*, 669–683.

Ickes, W., Schermer, B. J., & Steeno, J. (1979). Sex and sexrole influences in same-sex dyads. *Social Psychology Quarterly, 42*, 373–385.

Imada, A. S., & Hakel, M. D. (1977). Influence of nonverbal communication and rater proximity on impressions and decisions in simulated employment interviews. *Journal of Applied Psychology, 3*, 295–300.

Inman, W. S. (1935). Emotional factor in causation of diseases of the eye. *Transactions of the opthalmological society of the United Kingdom, 55*, 423–434.

Inman, W. S. (1955) 9-monthly scleritis in childless women. *British Journal of Medical Psychology, 28*, 177–182.

Irwin, B. L. (1971). Play therapy for a regressed schizophrenic patient. *Journal of Psychiatric Nursing and Mental Health Services, 9*, 30–32.

Isaacson, R. L., Douglas, R. J., Luber, J. F., & Schmaltz, L. (1971). *A primer of physiological psychology*. New York: Harper & Row.

Isenhart, M. (1978). An investigation of differential ability in decoding nonverbal cues of emotional meaning. (Doctoral dissertation, University of Denver.) *Dissertation Abstracts International, 39*, 4-a, 1925. (University Microfilms No. 78-18651.)

Jacobson, E. (1938). *Progressive Relaxation*. Chicago: University of Chicago Press.

Jaffe, J., Stern, P. N., & Perry, J. C. (1973). "Conversational" coupling of gaze behavior in prelinguistic human development. *Journal of Psycholinguistic Research, 2*, 321–330.

Janik, S. W., Wellens, A. R., Goldberg, M. L., & Dell'Osso, L. F. (1978). Eyes as the center of focus in the visual examination of human faces. *Perceptual and Motor Skills, 3*, 857–858.

Janisse, M. P., & Peavler, W. S. (1974). Pupillary response today: Emotion in the eye. *Psychology Today, 7*, 60–64.

JEB (Joan E. Biren). (1979). *Eye to eye: Portraits of lesbians*. Washington, DC: Glad Hag Books.

Jerison, H. J. (1973). *Evolution of the brain and intelligence*. New York: Academic Press.

*The Jerusalem Bible*. (A. Jones, Ed.). (1966). Garden City, NY: Doubleday.

Jobes, G. (1961). *Dictionary of mythology, folklore and symbols*. Metuchen, NJ: Scarecrow Press.

Johnson, B., & Boyd, T. (1978). The Eye Goddess and the evil eye. *Lady-Unique-Inclination-of-the-Night*, Cycle 3, 22–27.

Johnson, K. R. (1971). Black kinesics—some nonverbal communication patterns in the black culture. *Florida FL Reporter*, (Spring/Fall), 17–20, 57.

Jones, E. E., & Nisbett, R. E. (1971). The actor and the observer: Divergent perceptions of the causes of behavior. In E. E. Jones, D. E. Kanouse, H. H. Kelley, R. E. Nisbett, F. Valins, & B. Weiner (Eds.), *Attribution: Perceiving the causes of behavior*. New York: General Learning Press.

Jourard, S. M. (1971). *Self-disclosure: An experimental analysis of the transparent self*. New York: Wiley-Interscience.

Jourard, S. M., & Friedman, R. (1970). Experimenter-subject "distance" and self-disclosure. *Journal of Personality and Social Psychology*, 15, 278–282.

Joyce, J. (1916). *A portrait of the artist as a young man*. New York: Huebsch.

Jung, C. G. (1963). *Memories, dreams, reflections*. (A. Jaffe, Ed., and R. Winston & C. Winston, trans.). New York: Vintage Books.

Jurich, A. P., & Jurich, J. A. (1974). Correlations among nonverbal expressions of anxiety. *Psychological Reports*, 34, 199–204.

Kaila, E. (1932). Die Reaktionen des Sanglings auf das menschiche Gesicht. *Annales Universitatis Aboensis*, 17. Cited in R. A. Spitz & K. M. Wolf. (1946). The smiling response: A contribution to the ontogenesis of social relations. *Genetic Psychology Monographs*, 34, 57–125.

Kaiser, H. (1965). *Effective psychotherapy: The contribution of Helmuth Kaiser*. (L. B. Fireman, Ed.). New York: Free Press.

Kaplan, A., & Sedney, M. A. (1980). *Psychology and sex roles: An androgynous perspective*. Boston, MA: Little, Brown.

Karlen, A. (1971). *Sexuality and homo-sexuality: A new view*. New York: W. W. Norton.

Katz, R. (1963). *Empathy, it's nature and uses*. New York: Free Press of Glencoe.

Katz, R. (1965). *Body language: A study in unintentional communication*. (Doctoral dissertation, Harvard University.) *American Doctoral Dissertations*, 1965, p. 202.

Kearns, T. D. (1976). Situational and dispositional determination of nonverbal immediacy and relaxation. (Doctoral dissertation, Southern Illinois University, 1975.) *Dissertation Abstracts International*, 36, 6386. (University Microfilms No. 76-13, 253, 99.)

Keith, A. B. (1917). *The mythology of all races*, Vol. VI: *Indian*. Boston, MA: Marshall Jones.

Keller. (1960). Ablendlied. In *Motirgleiche Gedichte* (p. 37). Bamberg: Bayerische Verlagsanstadt.

Kelly, F. D. (1972). Communicational significance of therapist proxemic cues. *Journal of Consulting and Clinical Psychology*, 39, 345.

Kendon, A. (1967). Some functions of gaze-direction in social interaction. *Acta Psychologica*, 26, 22–63.

Kendon, A. (1978). Looking in conversation and the regulation of turns at talk: A comment on the papers of G. Beattie and D. R. Rutter et al. *British Journal of Social and Clinical Psychology*, 17, 23–24.

Kendon, A., & Cook, M. (1969). The consistency of gaze patterns in social interaction. *British Journal of Psychology*, 60, 481–494.

Kendon, A., & Ferber, A. (1973). A description of some human greetings. In P. M. Michael & J. H. Cook (Eds.), *Comparative ecology and behavior of primates* (pp. 591–668). London: Academic Press.

Kilpatrick, T. (1978, January 1). Watch it, kids! This teacher's got your number. *Washington Post*, F4.

Kimble, C. E., Forte, R. A., & Yoshikawa, J. C. (1981). Nonverbal concomitants of enacted emotional intensity and positivity: Visual and vocal behavior. *Journal of Personality*, 49, 271–283.

Kimble, C. E., & Olszewski, D. A. (1980). Gaze and emotional expressions. The effects of message positivity-negativity and emotional intensity. *Journal of Research in Personality*, 14, 60–69.

Kinross, J. (1964). *Ataturk*. London: Widenfeld & Nicholson.

Kirkland, J., & Lewis, C. (1976). Glance, look, gaze, and stare: A vocabulary for eye-fixation research. *Perceptual and Motor Skills*, 43, 1278.

Kirkpatrick, I. (1964). *Mussolini: Study in power*. New York: Hawthorne Books.

Kleck, R. E. (1968). Physical stigma and nonverbal cues emitted in face to face interaction. *Human Relations, 21*, 19-28.

Kleck, R. E., & Nuessle, W. (1968). Congruence between the indicative communicative functions of eye contact in interpersonal relations. *British Journal of Social and Clinical Psychology, 7*, 241-246.

Kleck, R. E., Ono, H., & Hastorf, A. H. (1966). The effects of physical deviance upon face-to-face interaction. *Human Relations, 19*, 425-436.

Klein, M., Heinmann, P., Isaacs, S., & Riviere, J. (1952). *Developments in psychoanalysis.* London: Hogarth.

Kleinke, C. L. (1972). Interpersonal attraction as it relates to gaze and distance between people. *Representative Research in Social Psychology, 3*, 105-120.

Kleinke, C. L. (1975). *First impressions: The psychology of encountering others.* Englewood Cliffs, NJ: Prentice-Hall.

Kleinke, C. L. (1977). Compliance to requests made by gazing and touching experimenters in field settings. *Journal of Experimental Social Psychology, 13*, 218-223.

Kleinke, C. L. (1980). Interaction between gaze and legitimacy of requesting compliance in a field study. *Journal of Nonverbal Behavior, 5*, 3-12.

Kleinke, C. L., Bustos, A. A., Meeker, F. B., & Staneski, R. A. (1973). Effects of self-attributed gaze on interpersonal evaluations between males and females. *Journal of Experimental Social Psychology, 9*, 154-163.

Kleinke, C. L., Desautels, M. J., & Knapp, B. E. (1977). Adult gaze and affective and visual responses of preschool children. *Journal of Genetic Psychology, 131*, 321-322.

Kleinke, C. L., & Pohlen, P. D. (1971). Affective and emotional responses as a function of other person's gaze and cooperativeness in a two-person game. *Journal of Personality and Social Psychology, 17*, 301-313.

Kleinke, C. L., & Singer, D. A. (1976). *Influence of gaze on compliance demanding and conciliatory requests in a field setting.* (Paper presented at the meeting of the Western Psychological Association, Los Angeles.)

Kleinke, C. L., Staneski, R. A., & Berger, D. E. (1975). Evaluation of an interviewer as a function of interviewer gaze, reinforcement of subject gaze, and interviewer attractiveness. *Journal of Personality and Social Psychology, 31*, 115-122.

Kleinke, C. L., Staneski, R. A., & Pipp, S. L. (1975). Effects of gaze, distance, and attractiveness on males' first impressions of females. *Representative Research in Social Psychology, 6*, 7-12.

Knapp, M. L. (1972). *Nonverbal communication in human interaction.* New York: Holt, Rinehart & Winston.

Knapp, M. L., Hart, R. P., Frederich, E. M., & Schulman, E. M. (1973). The rhetoric of goodbye: Verbal and nonverbal correlative human leavetaking. *Speech monographs, 40*, 182-198.

Knapp, M. L., Wiemann, J. M., & Daly, J. A. (1978). Nonverbal communication: Issues and appraisal. *Human Communication Research, 3*, 271-280.

Knight, D. J., Langmeyer, D., & Lundgren, D. C. (1973). Eye contact, distance and affiliation: The role of observer bias. *Sociometry, 36*, 390-401.

Koenig, O. (1975). *Urmotiv Auge: neuentdeckte Grundzuge menschl.* München, Zurich: Piper.

Kohler, I. (1963). The formation and transformation of the perceptual world. *Psychological Issues, 3*, Monogr. 12.

Kolvin, I. (1972). Infantile autism or infantile psychoses. *British Medical Journal, 5829*, 753-755.

Konia, C. (1970). A case of passive-feminine schizophrenia. *Journal of Orgonomy, 8*, 155-163.

Konia, C. (1977). The masochistic schizophrenic. *Journal of Orgonomy*, 11, 49–56.

Kramer, H., & Sprenger, J. (1971). *The malleus maleficarum*. (M. Summers, trans.) New York: Dover. (Originally published in 1448 under a Bull of Pope Innocent VIII.)

Krasner, L. (1961). Studies in the conditioning of verbal behavior. In S. Saporta (Ed.), *Psycholinguistics*. New York: Holt, Rinehart & Winston.

Kroll, J. (1978). Under the rainbow. *Newsweek*, 18, 89–90.

Kruger, L., & Stein, B. (1973). Primordial sense organs and the evolution of sensory systems. In E. C. Carterette & M. P. Friedman (Eds.), *Handbook of Perception*, Vol. III: *Biology of Perceptual Systems* (pp. 63–88). New York: Academic Press.

Kuh, K. (1971). *The open eye: In pursuit of art* (pp. 151–156). New York: Harper & Row.

Kummer, H. (1967). Tripartite relations in hamadryas baboons. In S. A. Altmann (Ed.), *Social communication among primates* (pp. 63–71). Chicago, IL: University of Chicago Press.

Kundu, M. R. (1976). Visual literacy: Teaching non-verbal communication through television. *Educational Technology*, 8, 31–33.

LaFrance, M., & Carmen, B. (1980). The nonverbal display of psychological androgyny. *Journal of Personality and Social Psychology*, 38, 36–49.

LaFrance, M., & Mayo, C. (1975). *Communication breakdown: The role of nonverbal behavior in interracial encounters*. (Paper presented at the International Communication Association Convention.)

LaFrance, M., & Mayo, C. (1976). Racial differences in gaze behavior during conversations: Two systematic observational studies. *Journal of Personality and Psychology*, 33, 547–52.

LaFrance, M., & Mayo, C. (1978). *Moving bodies: Nonverbal communication in social relationships*. Monterey, CA: Brooks/Cole Publishing.

Laing, R. D. (1960). *The divided self: A study of sanity and madness*. New York: Quadrangle.

Lamb, W., & Turner, D. (1969). *Management behaviour*. New York: International Universities Press.

Lambers, B. R. (1981). Eye opening of the newborn at and up to 20 minutes after birth. *Journal of Advanced Nursing*, 6, 455–459.

Lambert, W. W., & Lambert, W. E. (1964). *Social psychology*. Englewood Cliffs, NJ: Prentice-Hall.

Lamke, L., & Bell, N. (1981). Sex role orientation and relationship development in same-sex dyads (unpublished manuscript).

Landstrom, B. (1969). *The ship*. Garden City, NY: Doubleday.

Langdon, S. H. (1931). *The mythology of all races*, Vol. V: *Semitic*. Boston, MA: Marshall Jones.

Langer, E. J., Fiske, S., Taylor, S. E., & Chanowitz, B. (1976). Stigma, staring, and discomfort: A novel-stimulus hypothesis. *Journal of Experimental Social Psychology*, 12, 451–463.

Lassen, C. L. (1973). Effect of proximity on anxiety and communication in the initial psychiatric interview. *Journal of Abnormal Psychology*, 81, 226–232.

Lavater, L. C. (1953). *Essays on physiognomy: Designs to promote the knowledge and love of mankind*. (T. Halcroft, trans.) Farnborough, England: Gregg International.

Lawless, W. J. (1976). The role of dominance in human responses to staring. (Doctoral dissertation, Louisiana State University and Agricultural and Mechanical College.) *Dissertation Abstracts International*, 37, 3154-B.

Leavitt, L. A., & Donovan, N. L. (1979). Perceived infant temperament, locus of control, and maternal physiological response to infant gaze. *Journal of Research in Personality*, 13, 267–278.

LeCompte, W. F., & Rosenfeld, H. M. (1971). Effects of minimal eye-contact in the instruction period on impressions of the experimenter. *Journal of Experimental Social Psychology, 7*, 211–220.

LeCron, L. M., & Bordeaux, J. (1947). *Hypnotism today*. New York: Grune & Stratton.

Lefcourt, H. M., Rosenberg, F., Buckspan, B., & Steffy, R. A. (1967). Visual interaction and performance of process and reactive schizophrenics as a function of examiner's sex. *Journal of Personality, 35*, 535–546.

Lemon, L. T. (1968). A portrait: Motif as motivation and structure. In W. M. Schutte (Ed.), *Twentieth century interpretations of "A Portrait of the Artist as a Young Man."* Englewood Cliffs, NJ: Prentice-Hall.

LeSieg, T. (1968). *The eye book*. New York: Random House.

Levin, P. (1974). *Becoming the way we are: A transactional guide to personal development*. Berkeley, CA: Transactional Publications.

Levine, M. H. (1972). The effects of age, sex, and task on visual behavior during dyadic interaction. (Doctoral dissertation, Columbia University.) *Dissertation Abstracts International, 33*, 2325B–2326B. (University Microfilms No. 72-28, 063.)

Levine, M. H., & Sutton-Smith, B. (1973). Effects of age, sex, and task on visual behavior during dyadic interaction. *Developmental Psychology, 9*, 400–405.

Levy, T. M. (1973). Gaze behavior, repression-sensitization and social approach and avoidance. *Dissertation Abstracts International, 33*, 4544–4545.

Lewin, K. (1936). Principles of topological psychology. New York: McGraw-Hill.

Lewis, M., & Rosenbaum, L. (Eds.). (1974). *The effect of the infant on it's caretaker*. New York: Wiley.

Libby, W. L., Jr. (1970). Eye contact and direction of looking as stable individual differences. *Journal of Experimental Research in Personality, 4*, 303–312.

Libby, W. L., Jr., & Yaklevich, D. (1973). Personality determinants of eye contact and direction of gaze aversion. *Journal of Personality and Social Psychology, 27*, 197–206.

Lipps, T. (1913). *Zur Einfuehlung*. Leipzig: W. Englemann.

Little, R. B. (1965). Personal space. *Journal of Experimental Social Psychology, 1*, 237–247.

Little, W. A. (1961). *The eye complex in the dramas of Franz Grillparzer*. Ann Arbor, MI: University Microfilms No. 62-2762.

Lorenz, K. (1967). *On aggression* (M. K. Wilson, trans.). New York: Bantam Books.

Lott, B. (1981). A feminist critique of androgyny: Toward the elimination of gender attributions for learned behavior. In C. Mayo & N. M. Henley (Eds.), *Gender and nonverbal behavior*. New York: Springer-Verlag.

Lott, D. F., & Sommer, R. (1967). Seating arrangements and status. *Journal of Personality and Social Psychology, 7*, 90–95.

Lowen, A. (1975). *Bioenergetics*. New York: Coward, McCann & Geoghegan.

Lubow, A., & Copeland, J. B. (1978, July 10). How to get tough. *Newsweek*, 69.

Lumsden, E. A. Person perception as a function of the duration of the visual axes of the object person (unpublished paper).

MacGregor, F. C. (1974). *Transformation and identity: The face and plastic surgery*. New York: Quadrangle.

Machotka, P. (1965). Body movement as communication. In *Dialogue: Behavioral science research*. Boulder, CO: Western Interstate Commission for Higher Education.

Malone, J. (1975). Personal communication, March 1973.

Maloney, C. (Ed.) (1976). *The evil eye* (pp. 291–294). New York: Columbia University Press.

Mander, J. (1978, January). TV's capture of the mind. *Mother Jones*, 51–52, 56–61.

Mann, I., & Pirie, A. (1962). *The science of seeing*. Perth, Australia: Paterson Press.

Man's silent signals. (1969, June 13). *Time*, 86.

Mansfield, S. (1978, December 22). Jury will resume deliberation today in VA rape case. *Washington Post*, B8.

Mar, T. T. (1974). *Face reading: The Chinese art of physiognomy*. New York: Dodd, Mead & Co.

Marcelle, Y. M. (1977). Eye contact as a function of race, sex and distance. (Doctoral dissertation, Kansas State University.) *Dissertation Abstracts International, 37,* 4238A. (University Microfilms No. 76-30, 010.)

Markel, N. N. (1969). Relationship between voice-quality profiles and MMPI profiles in psychiatric patients. *Journal of Abnormal Psychology, 74,* 61–66.

Marks, H. E. (1971). The relationship of eye contact to congruence and empathy. *Dissertation Abstracts International, 32,* 1219.

Marler, P. (1965). Communication in apes and monkeys. In I. DeVore (Ed.), *Primate behavior* (pp. 571–584). New York: Holt, Rinehart & Winston.

Marshall, E. (1983). *Eye language: Understanding the eloquent eye.* Toronto: New Trend.

Marti-Ibanez, F. (1962). *Ariel: Essays on the arts and the history and philosophy of medicine.* New York: M. D. Publications.

Martin, J. (1981, January 1). Miss Manners: The eyes have it. *The Washington Post,* p. B-5.

Martin, M. (1980). *Gertrude Stein Gertrude Stein Gertrude Stein, a one-character play.* New York: Random House.

Matsusaka, K. (1972). On normalcy of mental traits in the infancy (1)-An observation of the mentally disturbed children. *Bulletin of the Faculty of Education, MIE University, 23,* 19–25.

Maurer, A. (1969). Peek-a-boo: An entry into the world of the autistic child. *Journal of Special Education, 3,* 309–312.

Maxwell, J. (1978, January). What your eyes tell you about your health. *Esquire,* 54–57.

May, R. (1976). *The courage to create.* New York: Bantam.

Mayo, C., & Henley, N. M. (Eds.). (1981). *Gender and Nonverbal Behavior.* New York: Springer-Verlag.

Mayo, C., & LaFrance, M. (1973). Gaze direction in interracial dyadic communication. (Paper presented at the meeting of the Eastern Psychological Association, Washington, DC.)

McBride, G., King, M. G., & James, J. W. (1965). Social proximity effects of galvanic skin response in adult humans. *Journal of Psychology, 61,* 153–157.

McConnell, O. L. (1967). Control of eye contact in an autistic child. *Journal of Child Psychology and Psychiatry and Allied Disciplines, 8,* 249–255.

McCulloch, W. S. (1951). Brain and behavior. In W. Dennis (Ed.), *Trends in psychological theory.* Pittsburgh: University of Pittsburgh Press.

McDowell, K. V. (1973). Accommodations of verbal and nonverbal behaviors as a function of the manipulation of interaction distance and eye contact. *Proceedings of the 81st Annual Convention of the American Psychological Association, 8,* 207–208.

McGinnis, J. (1930). Eye movements and optic nysagmus in early infancy. *Genetic Psychology Monographs, 8,* 321–430.

McGinnis, J. (1969). *The Selling of the President, 1968.* New York: Trident Press.

McLuhan, M. (1964). *Understanding media: The extensions of man.* New York: New American Library.

McPherson, M. (1980, January 6). Gov. Jerry Brown and the third-man-out-theory. *Washington Post,* G1, G3.

Meares, A. (1960). *A system of medical hypnosis.* New York: Julian Press.

Mehrabian, A. (1966). Immediacy: An indicator of attitudes in linguistic communication. *Journal of Personality, 34,* 26–34.

Mehrabian, A. (1967). Orientation behaviors and nonverbal attitude communication. *Journal of Communication, 17*, 324–332.

Mehrabian, A. (1968a). Relationship of attitude to seated posture, orientation, and distance. *Journal of Personality and Social Psychology, 10*, 26–33.

Mehrabian, A. (1968b). Communication without words. *Psychology Today, 4*, 53–55.

Mehrabian, A. (1969a). Significance of posture and position in the communication of attitude and status relationships. *Psychological Bulletin, 71*, 359–372.

Mehrabian, A. (1969b). Some referents and measures of nonverbal behavior. *Behavior Research Methods and Instrumentation, 1*, 203–207.

Mehrabian, A. (1970). A semantic space for nonverbal behavior. *Journal of Consulting and Clinical Psychology, 35*, 248–257.

Mehrabian, A. (1972). *Nonverbal communication*. New York: Adine-Atherton.

Mehrabian, A., & Ferris, S. R. (1967). Inference of attitudes from nonverbal communication in two channels. *Journal of Consulting and Clinical Psychology, 31*, 248–252.

Mehrabian, A., & Friar, J. T. (1969). Encoding of attitude by a seated communicator via posture and position cues. *Journal of Consulting and Clinical Psychology, 33*, 330–336.

Mehrabian, A., & Williams, M. (1969). Nonverbal concomitants of perceived and intended persuasiveness. *Journal of Personality and Social Psychology, 13*, 37–58.

Meili, R. (1957). *Anfange der Charakterentwicklung*. Stuttgart: Huber.

Melbin, M. (1972). *Alone and with others: A grammar of interpersonal behavior*. New York: Harper & Row.

Metzner, R., & Leary, T. (1967). On programming psychedelic experiences. *Psychedelic Review, 9*, 4–19.

Meynell, A. (1947). Eyes. In F. Page (Ed.), *Alice Meynell's prose and poetry* (pp. 222–226). London: Jonathan Cope.

Middleton, J. (1964). *Lugbara religion*. New York: Oxford University Press.

Midelfort, H. C. (1972). *Witch hunting in southwestern Germany, 1562–1684: The social and intellectual foundations*. Stanford, CA: Stanford University Press.

Miller, H. (1960). Hans Reichel. In L. Durrell (Ed.), *The Best of Henry Miller*. London: Heinemann.

Miller, J. E. (1968). Boats against the current. In Ernest H. Lockridge (Ed.), *Twentieth century interpretation of "The Great Gatsby."* Englewood Cliffs, NJ: Prentice-Hall.

Miller, J. G. (1961). Sensory overloading. In B. E. Flaherty (Ed.), *Psycho-physiological aspects of space flight*. New York: Columbia University Press.

Millet, K. (1977). Forward. In C. MacAdams, *Emergence*. New York: Chelsea House.

Minney, R. J. (1973). *Rasputin*. New York: McKay.

Mobbs, N. A. (1968). Eye-contact in relation to social introversion/extroversion. *British Journal of Social and Clinical Psychology, 7*, 305–306.

Monneret, S. (1978). Criticism and life; Notes on plates. In C. Esteban, J. Rudel, and S. Monneret, *Rembrandt*. New York: Excalibur Books.

Montagu, A. (1971). *The elephant man*. New York: Ballantine.

Monty, R. A., & Senders, J. W. (Eds.). (1976). *Eye movements and psychological processes*. Hillsdale, NJ: Lawrence Erlbaum Associates.

Morris, D. (1967). *The naked ape*. New York: McGraw-Hill.

Morris, D. (1977). *Manwatching: A field guide to human behavior*. New York: Abrams.

Morrison, R. (1969, March 15). Stars in your eyes, but not love. *Morning Herald,* Durham, NC, 17A.

Morrison, T. (1970). *The bluest eye*. New York: Holt, Rinehart, & Winston.

Moss, H. A. (1974). Communication in mother-infant interaction. In L. Krames,

P. Pliner, & T. Alloway (Eds.), *Advances in the study of communication and affect* (Vol. 1, pp. 171–191). New York: Plenum.

Moss, H. A., & Robson, K. S. (1968). Maternal influence in early social visual behavior. *Child Development, 39,* 401–408.

Moss, H. A., Robson, K. S., & Pederson, F. (1969). Determinants of maternal stimulation of infants and consequences of treatment for later reactions to strangers. *Developmental Psychology, 1,* 239–246.

Moss, L., & Cappannari, S. C. (1976). Mal'occhio, ayin ha ra, oculus fascinus, Judenblick: The evil eye hovers above. In C. Maloney (Ed.), *The evil eye.* New York: Columbia University Press.

Mueller, C., & Rudolph, M. (1966). *Light and vision.* New York: Time-Life Books.

Murphy, R. (1964). Social distance and the veil. *American Anthropologist, 66,* 1257–1274.

Nabokov, V. (1965). *The eye.* New York: Phaedra.

Nachshon, I., & Wapner, S. (1967). Effect of eye contact and physiognomy on perceived location of other person. *Journal of Personality and Social Psychology, 7,* 82–89.

Naha, E. (1975). *Horrors.* New York: Avon Books.

Natale, M. (1977). Induction of mood states and their effect of gaze behaviors. *Journal of Consulting and Clinical Psychology, 45,* 960.

Nelson, A. (1976). Mask-like faces as a presenting sign. *Journal of Orgonomy, 10,* 216–220.

Neumann, E. (1974). *The great mother.* Princeton, NJ: Princeton University Press.

Nevill, D. (1974). Experimental manipulation of dependency motivation and its effects on eye contact and measures of field dependency. *Journal of Personality and Social Psychology, 29,* 72–79.

Ney, P. G. (1973). Effect of contingent and non-contingent reinforcement on the behavior of an autistic child. *Journal of Autism and Childhood Schizophrenia, 3,* 115–127.

Nichols, K. A., & Champness, B. G. (1971). Eye gaze and the GSR. *Journal of Experimental Social Psychology, 7,* 623–626.

Nielsen, G. (1962). *Studies in self-confrontation.* Copenhagen: University of Copenhagen Psychological Laboratory.

Nierenberg, G. I. (1968). *The art of negotiation.* New York: Hawthorne Books.

Nierenberg, G. I., & Calero, H. H. (1971). *How to read a person like a book.* New York: Cornerstone Library.

Nin, A. (1967). *The diary of Anais Nin,* Vol. 2 (1934–1939). (G. Stuhlmann, ed.) New York & London: The Swallow Press & Harcourt Brace Jovanovich.

NineCurt, J. (1976a). *Nonverbal communication.* Cambridge, MA: National Assessment and Dissemination Center.

NineCurt, J. (1976b). *Teacher training pack for a course on cultural awareness.* Cambridge, MA: National Assessment and Dissemination Center.

NineCurt, J. (1980). *Ethnic body politics.* (Paper presented to the Institute for Nonverbal Communication Research Second Annual Research Conference.)

Noesjirwan, J. (1978). A laboratory study of proxemic patterns of Indonesians and Australians. *British Journal of Social Clinical Psychology, 17,* 333–334.

Nummenmaa, T. (1964). *The language of the face.* University of Jyuaskyla Studies in Education, Psychology and Social Research, No. 9.

O'Connor, L. (1970). Male dominance: The nitty gritty of oppression. *It Ain't Me Babe, 1,* 9–11.

O'Connor, N., & Hermelin, B. (1967). The selective visual attention of psychotic children. *Journal of Child Psychology and Psychiatry and Allied Disciplines, 8,* 167–179.

*Off our backs*. (1981, May). Special issue: Women with disabilities.

O'Riordain, S., & Daniel, G. (1964). *New grange and the bend of the Boyne*. New York: Praeger.

Orne, M. (1962). On the social psychology of the psychological experiment with particular reference to demand characteristics and their implications. *American Psychologist*, 17, 776–787.

Ornitz, E. M., & Ritvo, E. R. (1968). Perceptual inconsistency in early infantile autism. *Archives of General Psychiatry*, 18, 76–98.

Ortega y Gasset, J. (1957). *Man and people*. (W. R. Transk, trans.). New York: Norton.

Orwell, G. (1963). *1984*. New York: Harcourt, Brace and World.

Ousby, I. (1977). The broken glass: Vision and comprehension in "Bleak House." In C. Dickens, *Bleak House* (*An authoritative and annotated text.*) (G. Ford & S. Monad, Eds.). New York: W. W. Norton. (Original work published 1852).

Owens, R. (1973). "Dance and eye me (wicked)ly my breath a fixed sphere." In F. Howe & E. Bass (Eds.), *No more masks! An anthology of poetry by women*. Garden City, NY: Anchor Press.

Padden, C. (1976). "The eyes have it: Linguistic functions of the eyes in American sign language." In *Proceedings of the Second Gallaudet Symposium on Research in Deafness* (pp. 31–35). Washington, DC: Gallaudet Press.

Page, S. (1971). Social interaction and experimenter effects in the verbal conditioning experiment. *Canadian Journal of Psychology*, 25, 463–475.

Palmer, R. D. (1966). Visual acuity and excitement. *Psychosomatic Medicine*, 28, 364–374.

Palmer, R. D. (1970). Visual acuity and stimulus seeking behavior. *Psychosomatic Medicine*, 3, 277–284.

Papalia, D. E., & Olds, S. W. (1968). *Human development*. New York: McGraw-Hill.

Paradis, M. H. (1972). The effects of eye contact on positive and negative self-disclosure. *Dissertation Abstracts International*, 33, 2795.

Paramananda, S. (1977). *Book of daily thoughts and prayers*. Cohassett, MA: Vendanta Centre Publishers.

Patterson, M. L. (1973). Stability of nonverbal immediacy behaviors. *Journal of Experimental Social Psychology*, 9, 97–109.

Patterson, M. L. (1976). An arousal model of interpersonal intimacy. *Psychological Review*, 83, 235–245.

Patterson, M. L. (1977). Interpersonal distance, affect and Equilibrium Theory. *Journal of Social Psychology*, 101, 205–214.

Patterson, M. L. (1978a). Arousal change and cognitive labeling: Pursuing the mediators of intimacy exchange. *Environmental Psychology and Nonverbal Behavior*, 3, 17–22.

Patterson, M. L. (1978b). *Nonverbal intimacy exchange: Problems and prospects*. (Paper presented at the Second National Conference on Body Language, New York.)

Patterson, M. L., Jordan, A., Hogan, M., & Frerker, D. (1981). Effects of nonverbal intimacy on arousal and behavioral adjustment. *Journal of Nonverbal Behavior*, 3, 184–198.

Patterson, M. L., & Sechrest, L. B. (1970). Interpersonal distance and impression formation. *Journal of Personality*, 38, 161–166.

Payne, C. (1977). Cultural differences and their implications for teachers. *Integrated Education*, 15, 42–45.

Pedhazur, E. J., & Tetenbaum, T. J. (1979). Bem sex role inventory: A theoretical and methodological critique. *Journal of Personality and Social Psychology*, 37, 996–1016.

Pellegrini, R. J., Hicks, R. A., & Gordon, L. (1970). The effects of an approval-seeking induction on eye contact in dyads. *British Journal of Social and Clinical Psychology*, 9, 373–374.

Penner, L. A. (1971). Interpersonal attraction toward a black person as a function of value importance. *Personality*, 2, 175–187.

Perls, F. S. (1969). *Gestalt therapy verbatim*. Moab, UT: Real People Press.

Perls, L. (1973). Some aspects of gestalt therapy. (Paper presented to the Orthopsychiatric Association.)

Peto, A. (1969). Terrifying eyes. *Psychoanalytic Study of the Child*, 24, 197–212.

Pfeiffer, E. (1968). *Disordered behavior*. New York: Oxford University Press.

Pfeiffer, J. W. (1973). Conditions which hinder effective communication. In 1973 *Annual Handbook for Group Facilitators* (pp. 120–123). San Diego, CA: University Associates.

Polanyi, M. (1967). *The tacit dimension*. Garden City, NY: Doubleday.

Pollster, E., & Pollster, M. (1973). *Gestalt therapy integrated*. New York: Brunner/ Mazel.

Polyak, S. (1957). *The vertebrate visual system*. Chicago: University of Chicago Press.

Porter, N., & Geis, F. (1981). Women and nonverbal leadership cues: When seeing is not believing. In C. Mayo & N. M. Henley (Eds.), *Gender and Nonverbal Behavior*. New York: Springer-Verlag.

Ramsden, P. (1976). *Top team planning*. New York: Wiley.

Rank, O. (1913). Eine noch nicht Beschriebene form des Odipus-Traum. *Internationale Zeitschrift fur Psychoanalyse*, 1, 151–156.

Ratnavale, D. N. (1984). Eye contact—Windows of vulnerability in psychotherapeutic work. (A paper presented before the D.C. Institute of Mental Hygiene.)

Reece, M., & Whitman, R. (1962). Expressive movements, warmth and verbal reinforcement. *Journal of Abnormal Social Psychology*, 61, 234–236.

Reich, W. (1972). *Character analysis*, 3rd ed. (V. R. Carfagno, trans.). New York: Farrar, Straus & Giroux.

Reiss, E. (1969). The symbolic surface of the Canterbury Tales: The Monk's portrait (Part II). *Chaucer Review*, 3, 12–28.

Reitler, R. (1913). Beitrage zur Symbolik. *Internationale Zeitschrift für Psychoanalyse*, 1, 159–161.

Rendel, P. (1974). *Introduction to the chakras*. Wellingborough, England: Aquarian Press.

Reynolds, H. N. (1977). Personal communication.

Rheingold, H. L. (1961). The effect of environmental stimulation upon social and exploratory behavior in the human infant. In B. M. Foss (Ed.), *Determinants of infant behavior* (Vol. 1). New York: John Wiley.

Richer, J. M., & Cross, R. G. (1976). Gaze aversion in normal and autistic children. *Acta Psychiatrica Scandinavica*, 53, 193–210.

Riemer, M. D. (1949). The averted gaze. *Psychiatric Quarterly*, 23, 108–115.

Riemer, M. D. (1955). Abnormalities of the gaze: A classification. *Psychiatric Quarterly*, 29, 659–672.

Rime, B., & McCusker, L. (1976). Visual behaviour in social interaction: The validity of eye-contact assessments under different conditions of observation. *The British Journal of Psychology*, 67, 507–514.

Rimland, B. (1963). *Infantile autism*. New York: Appleton-Century-Crofts.

Roberts, J. M. (1976). Belief in the evil eye in world perspective. In Clarence Maloney (Ed.), *The evil eye*. New York: Columbia University Press.

Robertson, M. (1975). *A history of Greek art*. New York: Cambridge University Press.

Robin, M. (1980). Interaction process analysis of mothers with their newborn infants. *Early Child Development and Care*, 6, 93–108.

Robson, K. S. (1967). The role of eye-to-eye contact in maternal-infant attachment. *Journal of Child Psychology, Psychiatry and Allied Disciplines, 8,* 13–25.

Robson, K. S., Pederson, F. A., & Moss, H. A. (1969). Developmental observations of diadic gazing in relation to the fear of strangers and social approach in behavior. *Child Development, 40,* 619–627.

Rogers, C. R. (1957). The necessary and sufficient conditions of therapeutic personality change. *Journal of Consulting Psychology, 21,* 95–103.

Rogers, C. R. (1961). *On becoming a person.* Boston: Houghton Mifflin.

Rolleston, J. D. (1942). Communications: Opthalmic folk-lore. *British Journal of Opthalmology, 26,* 481–502.

Rosenberg, J. L. (1973). *Total orgasm.* New York: Random House.

Rosenberg, M. (1969). The conditions and consequences of evaluation apprehension. In R. Rosenthal & R. Rosnow (Eds.), *Artifact in behavior research.* New York: Academic Press.

Rosenthal, R. (1966). *Experimenter effects in behavioral research.* New York: Appleton-Century-Crofts.

Rosenthal, R., & DePaulo, B. M. (1979). Sex differences in accommodation nonverbal communication. In R. Rosenthal (Ed.), *Skill in nonverbal communication.* Cambridge, MA: Oelsgeschlager, Gunn, & Hain.

Rosetti, D. G. (1970). 97. A superscription. From "The house of life." In A. M. Eastman et al. (Eds.), *The Norton anthology of poetry.* New York: W. W. Norton.

Rouchouse, J. C. (1978). Human ethology and nonverbal communication among infants. *Enfance, 1,* 13–30.

Rousett, J. (1981). *Leurs yeux se rencontrerent.* Paris: Librarie Jose Corti.

Rubin, J. E. (1968). *Impression change as a function of level of self-disclosure.* (Master's thesis, University of Florida.)

Rubin, Z. (1970). Measurement of romantic love. *Journal of Personality and Social Psychology, 16,* 265–73.

Rubin, Z. (1973). *Liking and loving, an invitation to social psychology.* New York: Holt, Rinehart & Winston.

Ruesch, J. (1961). *Therapeutic communication.* New York: Norton.

Ruesch, J. (1972). *Semiotic approaches to human relations.* Paris: Mouton.

Ruesch, J., & Kees, W. (1956). *Nonverbal communication.* Berkeley, CA: University of California Press.

Russo, N. F. (1971). Eye contact and distance relations in children. *Dissertation Abstracts International, 31,* 7644–7645.

Russo, N. F. (1975). Eye contact, interpersonal distance and the equilibrium theory. *Journal of Personality and Social Psychology, 31,* 497–502.

Rutter, D. R. (1973). Visual interaction in psychiatric patients: A review. *British Journal of Psychiatry, 123,* 193–202.

Rutter, D. R. (1976). Visual interaction in recently admitted and chronic long-stay schizophrenic patients. *British Journal of Social and Clinical Psychology, 15,* 295–303.

Rutter, D. R. (1978). The timing of looks in dyadic conversation. *The British Journal of Social and Clinical Psychology, 17,* 17–21.

Rutter, D. R., Morley, I. E., & Graham, J. C. (1972). Visual interaction group of introverts and extroverts. *European Journal of Social Psychology, 2,* 371–384.

Rutter, D. R., & O'Brien, P. (1980). Social interaction in withdrawn and aggressive maladjusted girls: A study of gaze. *Journal of Child Psychology, Psychiatry and Allied Disciplines, 21,* 59–66.

Rutter, D. R., & Stephenson, G. M. (1972). Visual interaction in a group of schizophrenic and depressed patients. *British Journal of Social and Clinical Psychology, 11,* 57–62.

Rutter, D. R., & Stephenson, G. M. (1979). The role of visual communication in social interaction. *Current Anthropology, 1*, 124–125.

Samuels, M., & Samuels, N. (1975). *Seeing with the mind's eye: The history, techniques and uses of visualization.* New York: Random House.

Sanders, G. (1979, July 24). A time to look beyond eye contact. *New York Times.*

Sapir, E. (1927). The unconscious patterning of behavior in society. In E. S. Dummer (Ed.), *The unconscious: A symposium.* New York: Knopf.

Saposnek, D. T. (1972). An experimental study of rage-reduction treatment of autistic children. *Child Psychiatry and Human Development, 3*, 50–62.

Saraydarian, H. (1976). *The science of meditation.* Agoura, CA: The Aquarian Educational Group.

Sartre, J. P. (1963). *"No Exit" and "The Flies."* (S. Gilbert, trans.). New York: Knopf.

Sartre, J. P. (1964). *The age of reason* (E. Sutton, trans.). New York: Bantam Books.

Satir, V. (1976). *Making contact.* Millbrae, CA: Celestial Arts.

Schachtel, E. (1959). *Metamorphosis.* New York: Basic Books.

Scharff, B. (1978, May 3). "The eyes and hypnosis." (Unpublished paper, University of Maryland.)

Scheflen, A. E. (1963). Communication and regulation in therapy. *Psychiatry, 26*, 126–136.

Scheflen, A. E. (1964). The significance of posture in communication systems. *Psychiatry, 27*, 316–331.

Scheflen, A. E. (1966). Natural history method in psychotherapy: Communicational research. In L. A. Gottschaulk & A. H. Auerbach (Eds.), *Methods of research in psychotherapy.* New York: Appleton-Century-Crofts.

Scheflen, A. E. (1968). Human communication: Behavioral programs and their integration in interaction. *Behavioral Science, 13*, 44–55.

Scheflen, A. E. (1973). *Body language and the social order: Communication as behavioral control.* Englewood Cliffs, NJ: Prentice-Hall.

Scheflen, A. E. (1974). *How behavior means.* Garden City, NY: Anchor Press Doubleday.

Scheflen, A. E. (1978). Personal communication.

Scherer, S. E., & Schiff, M. R. (1973). Perceived intimacy, physical distance and eye contact. *Perceptual and Motor Skills, 36*, 835–884.

Scherwitz, L., & Helmreich, R. (1973). Interactive effects of eye contact and verbal content on interpersonal attraction in dyads. *Journal of Personality and Social Psychology, 25*, 6–14.

Scheutz, A. (1948). Sartre's theory of the alter ego. *Philosophy and Phenomenological Research, 9*, 181–199.

Schieffelin, B. B. (1977). Looking and talking: The functions of gaze direction in the conversations of a young child and her mother. (Unpublished manuscript.)

Schlaegel, T. F., & Hoyt, M. (1957). *Psychosomatic opthalmology.* Baltimore, MD: Williams & Wilkins.

Schmidt, W., & Hore, T. (1970). Some nonverbal aspects of communication between mother and preschool child. *Child Development, 41*, 889–896.

Schneider, D. J., Hastorf, A. H., & Ellsworth, P. C. (1979). *Person perception.* Reading, MA: Addison-Wesley.

Schneider, F. W., Coutts, L. M., & Garrett, W. A. (1977). Interpersonal gaze in a triad as a function of sex. *Perceptual and Motor Skills, 44*, 184.

Schooler, C., & Silverman, J. (1969). Perceptual styles and their correlates among schizophrenic patients. *Journal of Abnormal Psychology, 74*, 459–470.

Schopler, E. (1966). Visual versus tactual receptor preferences in normal vs. schizophrenic children. *Journal of Abnormal Psychology, 71*, 108–114.

Schutte, W. M. (1968). *Twentieth century interpretations of "A Portrait of the Artist as a Young Man."* Englewood Cliffs, NJ: Prentice-Hall.

Schutz, W. (1958). *FIRO: A three-dimensional theory of interpersonal behavior.* New York: Rinehart & Co.

Schwarz, L. M. (1982). Multichannel nonverbal communication: More signals do not always improve understanding. (Doctoral dissertation, Temple University.) *Dissertation Abstracts International, 43,* 0917B. (University Microfilms No. 82-17794.)

Seagull, A. A. (1967). The treatment of a suicide threat. *Psychotherapy: Theory, Research and Practice, 4,* 41-43.

Sebeok, T. A. (Ed.) (1968). Animal communication—Techniques of study and results of research. Bloomington, IN: Indiana University Press.

Serber, M. (1972). Teaching the nonverbal components of assertive training. *Journal of Behavior Therapy and Experimental Psychiatry, 3,* 179-183.

Shakespeare, W. (1919). *The complete works of Shakespeare.* (W. J. Craig, Ed.). New York: Oxford University Press.

Shapiro, E. (1977). Destructive cultism. *American Family Physician, 15,* 80-83.

Shapiro, T., & Stine, J. (1965). The figure drawings of three-year-old children. *Psychoanalytic Study of the Child, 20,* 298-309.

Shaw, M. E., Bowman, J. T., & Haemmerlie, F. M. (1971). The validity of measurement of eye-contact. *Educational and Psychological Measurement, 31,* 919-925.

Shenker, I. (1969, February 22). A new study: Behavior while talking. *New York Times,* B-1, C-77.

Sherer, M., & Rogers, R. W. (1980). Effects of therapists nonverbal communication on rated skill and effectiveness. *Journal of Clinical Psychology, 36,* 696-700.

Shirley, M. M. (1933). *The first two years* (Vols. I & II). Minneapolis: University of Minnesota Press.

Shorr, J. E. (1977). *Go see the movie in your head.* New York: Popular Library, CBS Publications.

Siegman, A. W., & Feldstein, S. (Eds.). (1978). *Nonverbal behavior and communication.* Hillsdale, NJ: Lawrence Erlbaum Associates.

Sieveking, N. A. (1972). Behavioral therapy: Bag of tricks or point of view? Treatment for homosexuality. *Psychotherapy: Theory, Research and Practice, 9,* 32-35.

Sigerson, D. (1920). The eyes. In *A dull day in London.* London: Eveleigh Nash.

Sill, E. R. (1900). Felt location of the "I." In *The Prose of Edward Rowland Sill.* New York: Houghton Mifflin.

Simmel, G. (1921). Sociology of the senses: Visual interaction. In R. Park & E. Burgess (Eds.), *Introduction to the science of sociology.* Chicago: University of Chicago Press.

Simmel, G. (1924). Sociology of the senses: Visual interaction. In R. Park & E. Burgess (Eds.), *Introduction to the science of sociology.* Chicago: University of Chicago Press.

Simon, J. (1977). Introduction. In *The screenplays of Lina Wertmuller.* New York: Quadrangle.

Smith, B. J., Sanford, F., & Goldman, M. (1977). Norm violations, sex, and the "blank stare." *Journal of Social Psychology, 103,* 49-55.

Smith, J., & Sheahan, M. (1979). Verbal and nonverbal interaction stages in human relationships. Paper presented to the annual convention of the Eastern Communication Association, Philadelphia, PA.

*Smithsonian Magazine.* (1979, March 10). Letters to the editor.

Smyth, J. (1977, December 5). A little intrigue around the eyes. *Washington Post,* B1, 4.

Smythe, R. H. (1975). *Vision in the animal world.* New York: St. Martin's Press.

Snyder, M., Grether, J., & Keller, K. (1974). Staring and compliance: A field experiment on hitchhiking. *Journal of Applied Social Psychology, 4,* 165-170.

Sobelman, S. A. (1973). The effects of verbal and nonverbal components on the judged level of counselor warmth. (Doctoral dissertation, American University.) *Dissertation Abstracts International, 35,* 273. (University Microfilms No. 74-14199.) 14199.)

Sommer, R. (1959). Studies in personal space. *Sociometry, 22,* 247-260.

Sommer, R. (1969). *Personal space, the behavioral basis of design.* Englewood Cliffs, NJ: Prentice-Hall.

Sophocles. (1958). *The Oedipus plays of Sophocles* (Paul Roche, trans.). New York: Mentor Books.

Sorensen, T. C. (1984). *A different kind of presidency: A proposal for breaking the political deadlock.* New York: Harper & Row.

Spaulding, R. (1978). Personal communication.

Spence, D. P., & Feinberg, C. (1967). Forms of defensive looking: A naturalistic experiment. *Journal of Nervous and Mental Disease, 145,* 261-271.

Sperry, R. W. (1951). Mechanisms of neural maturation. In S. S. Stevens (Ed.), *Handbook of experimental psychology.* New York: Wiley.

Spiegel, H. (1972). An eye-roll test for hypnotizability. *The American Journal of Clinical Hypnosis, 15,* 25-28.

Spiegel, H. (1974). The Grade 5 Syndrome: The highly hypnotizable person. *The International Journal of Clinical and Experimental Hypnosis, 22,* 303-319.

Spiegel, H. (1977). The Hypnotic Induction Profile (HIP): A review of its development. *Annals of the New York Academy of Sciences, 296,* 129-142.

Spiegel, H., Aronson, M., Fleiss, J. L., & Haber, J. (1976). Psychometric analysis of the Hypnotic Induction Profile. *International Journal of Clinical and Experimental Hypnosis, 24,* 300-315.

Spitz, R. A., & Wolf, K. M. (1946). The smiling response: A contribution to ontogenesis of social relations. *Genetic Psychology Monographs, 34,* 57-125.

Spitzer, R., & Mann, I. (1950). Congenital malformations in the teeth and eyes in mental defectives. *Journal of Mental Science, 96,* 681-709.

Spooner, B. (1976). The evil eye in the Middle East. In C. Maloney (Ed.), *The evil eye.* New York: Columbia University Press.

Spretnak, C. (1978). *Lost goddesses of early Greece. A collection of pre-Hellenic mythology.* Berkeley, CA: Moon Books.

Stanley, G., & Martin, D. S. (1968). Eye contact and the recall of material involving competitive and noncompetitive associations. *Psychonomic Science, 13,* 337-338.

Starkweather, C. W. (1977). Disorders of nonverbal communication. *Journal of Speech and Hearing Disorders, XLIII,* 535-546.

Starr (1979). *On stage.* Comic strip, Lehigh, PA.

Stass, J. W., & Willis, F. N. Jr. (1967). Eye contact, pupil dilation, and personal preference. *Psychonomic Science, 7,* 375-376.

Stechler, G., & Carpenter, G. (1967). A viewpoint on early affective development. In J. Hellmuth (Ed.), *The Exceptional Infant* (Vol. 1, pp. 163-189). Seattle: Special Child Publications.

Stechler, G., & Latz, E. (1966). Some observations on attention and arousal in the human infant. *Journal of the American Academy of Child Psychiatry, 5,* 517-525.

Steiner, C. (1971). Radical psychiatry: Principles. In J. Agel (Ed.), *The Radical Therapist.* New York: Ballantine Books.

Stephenson, G. M., & Rutter, D. R. (1970). Eye-contact, distance and affiliation reevaluation. *British Journal of Psychology, 61,* 385-393.

Stephenson, G. M., Rutter, D. R., & Dore, S. R. (1973). Visual interaction and distance. *British Journal of Psychology, 64,* 251–257.

Stern, D. (1974). Mother and infant at play. In M. Lewis and L. Rosenbaum (Eds.), *The effect of the infant on its caretaker.* New York: Wiley.

Stern, D., Beebe, B., Jaffe, J., & Benett, S. (1977). The infants' stimulus world during social interaction: A study of caregiver behaviors with particular reference to repetition and timing. In H. R. Schaffer (Ed.), Studies in mother-infant interaction. New York: Academic Press.

Stolz, S. B., & Wolf, M. M. (1969). Visually discriminated behavior in a "blind" adolescent retardate. *Journal of Applied Behavior Analysis, 2,* 65–77.

Stratton, G. M. (1934). Emotional reactions connected with differences in cephalic index, shade of hair, and the color of eyes in Caucasians. *American Journal of Psychology, 46,* 409–419.

Strongman, K. T., & Champness, B. G. (1968). Dominance hierarchies and conflict in eye contact. *Acta Psychologica, 28,* 376–386.

Strosberg, I. M., & Vics, I. I. (1962). Physiological changes in the eye during hypnosis. *American Journal of Clinical Hypnosis, 4,* 264–267.

Styron, W. (1967). *The confessions of Nat Turner.* New York: New American Library.

Sullivan, H. S. (1954). *The psychiatric interview.* New York: W. W. Norton.

Sumby, H., & Pollack, I. (1954). Visual contribution to speech intelligibility in noise. *Journal of the Accoustical Society of America, 26,* 212–215.

Sutton-Vane, S. (1958). *The story of the eyes.* New York: Viking Press.

Symonds, C. (1972, May 8). A vocabulary of sexual enticement and proposition. *Journal of Sex Research,* 136–139.

Szasz, T. (1973). The myth of mental illness. In P. Brown (Ed.), *Radical Psychology.* New York: Harper & Row.

Talbot, D. (1980). *The saturn myth.* Garden City, NY: Doubleday.

Tankard, J. W. (1970). The connotative meaning of the eye contact cue to a perceiver. (Doctoral dissertation, Stanford University, 1970.) *Dissertation Abstracts International,* 31/04-B. (University Microfilm No. 70-18486, 2330.)

Teacher finds lies in facial expression. (1977, September 25). *New York Times,* 20.

Tecce, J. J. (1978). Personal communication, May 20.

Tecce, J. J., and Cole, J. O. (1976). The distraction-arousal hypothesis, CNV, and schizophrenia. In D. I. Mostofsky (Ed.), *Behavior control and modification of physiological activity* (pp. 162–219). Englewood Cliffs, NJ: Prentice-Hall.

Tecce, J. J., Savignano-Bowman, J., and Cole, J. O. (1978). Drug effects on contingent negative variation and eyeblinks: The distraction-arousal hypothesis. In M. A. Lipton, A. DiMascio, & K. F. Killam (Eds.), *Psychopharmacology: A generation of progress.* New York: Raven Press, 745–757.

Teitelbaum, J. M. (1976). The leer and the loom—social controls on handloom weavers. In C. Maloney (Ed.), *The evil eye* (pp. 63–75). New York: Columbia University Press.

Tennov, D. (1980). *Love and limerance: The experience of being in love.* New York: Stein and Day.

Thayer, S., & Schiff, W. (1974). Observer judgment of social interaction: Eye contact and relationship inferences. *Journal of Personality and Social Psychology, 1,* 110–114.

Thayer, S., & Schiff, W. (1975). Eye contact, facial expression, and the experience of time. *Journal of Social Psychology, 95,* 117–124.

Thayer, S., & Schiff, W. (1977). Gazing patterns and attribution of sexual involvement. *Journal of Social Psychology, 101,* 235–246.

Thomas, E. (1968). Movements of the eye. *Scientific American, 21,* 88–95.

Thompkins, S. S., & McCarter, R. (1964). What and where are the primary effects? Some evidence for a theory. *Perceptual and Motor Skills, 18,* 119–158.

Thomson, D. S. (1975). *Human behavior: Language.* New York: Time-Life Books.

Tomkins, S. (1963). *Affect, imagery, and consciousness: Negative affects* (Vol. 2). New York: Springer Publishing.

Tourney, G., & Plazak, D. J. (1954). Evil eye in myth and schizophrenia. *Psychiatric Quarterly, 28,* 478–495.

Trachtenberg, J. (1974). *Jewish magic and superstition: A study in folk religion.* New York: Atheneum.

Tripp, R. M. (1967). *A study of cardiac afferent influence on an involuntary motor behavior: A determinant of blinking.* (Doctoral dissertation, Duke University.)

Trout, D. L., & Rosenfeld, H. M. (1980). The effect of postural lean and body congruence on the judgment of psycho-therapeutic rapport. *Journal of Nonverbal Behavior, 3,* 176–177.

Truax, C. B., & Wargo, D. G. (1966). Psychotherapeutic encounters that change behavior: For better or for worse. *American Journal of Psychotherapy, 20,* 499–520.

Truax, C. B., Wargo, D. G., Frank, J. D., Imber, S. D., Battle, C. C., Hoehn-Saric, R., Nash, E. H., & Stone, A. R. (1966). The therapist's contribution to accurate empathy, non-possessive warmth, and genuineness in psychotherapy. *Journal of Clinical Psychology, 22,* 331–334.

Tuchman, B. W. (1966). *The proud tower: A portrait of the world before the war, 1890–1914.* New York: Macmillan.

Tuchman, B. W. (1978). *A distant mirror: The calamitous fourteenth century.* New York: Alfred A. Knopf.

Unger, L. (Ed.). (1974). *American writers* (Vol. III, p. 24). New York: Charles Scribner.

Van der Drift, H. (1966). Het verschijnsel Blikgevoeligheid bij Schizofrenen. *Nederlands Tijdschrift voor Geneeskunde, 110,* 1128–1153.

Van Houten, R., Nau, P. A., MacKenzie-Keating, S. E., & Colanicchia, B. (1982). An analysis of some variables influencing the effectiveness of reprimands. *Journal of Applied Behavioral Analysis, 15,* 65–83.

Vaughn, B. E., & Waters, E. (1981). Attention structure, sociometric status, and dominance: Interrelations, behavioral correlates, and relationships to social competence. *Developmental Psychology, 17,* 275–288.

Villa-Lovoz, T. R. (1975). The effects of counselor eye contact, fluency, and addressing on client preference. (Doctoral dissertation, University of Wyoming, 1975). *Dissertation Abstracts International, 36/04A,* 2039A. (University Microfilms No. 75-22341.)

Vine, I. (1971). Judgment of direction of gaze—an interpretation of discrepant results. *British Journal of Clinical Psychology, 10,* 320–331.

Vine, I. (1973). The role of facial-visual signalling in early social development. In M. von Cranach & I. Vine (Eds.), *Social communication and movement.* New York and London: Academic Press.

Vlietstra, A. G., & Manske, S. H. (1981). Looks to adults preferences for adult males and females, and interpretations of an adult's gaze by preschool children. *Merrill-Palmer Quarterly, 27,* 31–41.

Wada, J. A. (1961). Modification of cortically induced responses in brain stem of shift of attention in monkeys. *Science, 133,* 40–42.

Wahl, P. (1969, January). The camera that can read your mind. *Science and Mechanics,* 38–39, 89.

Wallach, M. B., & Wallach, S. S. (1964). Involuntary eye movement in certain schizophrenics. *Archives of General Psychiatry, 11*, 71–73.

Wallach, S. S., Wallach, M. B., & Yessin, G. (1960). Observation of involuntary eye movements in certain schizophrenics. *Journal of Hillside Hospital, 9*, 224–227.

Walters, R. H., & Parke, R. D. (1965). The role of the distance receptors in the development of social responsiveness. In L. Lipsitt & C. Spiker (Eds.), *Advances in child development and behavior* (Vol. 2). New York: Academic Press.

Watson, O. M. (1970). *Proxemic behavior: A cross-cultural study.* The Hague: Mouton.

Watson, O. M., & Graves, T. D. (1966). Quantitative research in proxemic behavior. *American Anthropologist, 68*, 971–985.

Waxer, P. (1974). Nonverbal cues for depression. *Journal of Abnormal Psychology, 83*, 319–322.

Waxer, P. H. (1978). *Nonverbal aspects of psychotherapy.* New York: Praeger.

Webb, E. J., Campbell, D. T., Schwartz, R. D., & Sechrest, L. (1966). *Unobtrusive measures: Nonreactive research in the social sciences.* Chicago: Rand McNally.

Webbink, P. G. (1974). *Eye contact and intimacy.* (Doctoral dissertation, Duke University.)

Webbink, P. G. (1980). Eye and image—eye contact and guided imagery in psychotherapy. In J. E. Shorr, G. Sobel-Whittington, P. Robin, & J. A. Connella (Eds.), *Imagery.* New York: Plenum.

Webbink, P. G. (1981). Nonverbal behavior and lesbian/gay orientation. In C. Mayo & N. M. Henley (Eds.), *Gender and nonverbal behavior.* New York: Springer-Verlag.

Webbink, P. G. (unpublished). Eye contact and imagery in psychotherapy.

Weisbrod, R. M. (1967). Looking behavior in a discussion group. Unpublished paper, Cornell University. Cited in M. Argyle and A. Kendon, The experimental analysis of social performance. In L. Berkowitz (Ed.), *Advances in experimental social psychology.* New York: Academic Press.

Weitz, S. (1976). Sex differences in nonverbal communication. *Sex Roles, 2*, 175–184.

Weitz, S. (1979). *Nonverbal communication: Readings with commentary.* New York: Oxford University Press.

Weitzenhoffer, A. M. (1957). *General techniques of hypnotism.* New York: Grune & Stratton.

Welford, A. T. (1958). *Aging and human skill.* London and New York: Oxford University Press.

Werblowsky, R. J., & Wigoder, G. (1967). *The encyclopedia of the Jewish religion.* London: Phoenix House.

Wharton, E. (1910). The eyes. In *Tales of men and ghosts* (pp. 243–274). New York: Charles Scribner.

Whatmore, G. B., & Ellis, R. M. (1962). Further neurophysiologic aspects of depressed states: An electromyographic study. *Archives of General Psychology, 6*, 243–253.

Wheeler, R. W., Baron, J. C., Mitchell, S., & Ginsburg, H. J. (1979). Eye contact and the perception of intelligence. *Bulletin of the Psychonomic Society, 13*, 101–102.

White, T. (1984). Personal communication, October 16.

Whyte, W. F. (1949). The social structure of the restaurant. *American Journal of Sociology, 54*, 302–308.

Wichman, H. (1970). Effects of isolation and communication on co-operation in a two-person game. *Journal of Personality and Social Psychology, 16*, 114–120.

Wilbur, K. (1983). Eye to Eye: The quest for the new paradigm. Garden City, NY: Anchor Press/Doubleday.

Wilkinson, T. A. (1969). Eye manifestations of emotion. *Journal of American Ophthalmology Association, 40,* 516–517.

Williams, E. (1974). An analysis of gaze and schizophrenia. *British Journal of Social and Clinical Psychology, 13,* 1–8.

Willner, R. A. (1968). *Charismatic political leadership: A theory.* Research Monograph #32. Princeton: Princeton University Center of International Studies.

Wilson, D. C. (1974). *Bright eyes.* New York: McGraw-Hill.

Winick, C., & Holt, H. (1962). Eye and face movements as non-verbal communication in group therapy. *Journal of Hillside Hospital, 11,* 67–79.

Witherow, J. K. (1981, May). Hello walls. *Off our backs, xi,* 2–3.

Witkin, H. A. (1949). Sex differences in perception. *Transactions of the New York Academy of Sciences, 12,* 22–26.

Witkin, H. A. (1950). Individual differences in ease of perception of embedded figures. *Journal of Personality, 19,* 1–15.

Wolf, L. A. (1976). The effects of visibility and gender in dyadic interaction. (Doctoral dissertation, Yale University.) *Dissertation Abstracts International, 37B,* 534. (University Microfilms No. 76-14575.)

Wolff, P. H. (1963). Observations on the early development of smiling. In B. M. Foss (Ed.), Determinants of infant behavior, Vol. II. London: Methuen.

Wolpe, J. (1958). *Psychotherapy by reciprocal inhibition.* Stanford, CA: Stanford University Press.

Wood, L. A., & Saunders, J. (1962). Blinking frequency: A neurophysiological measurement of psychological stress. *Diseases of the Nervous System, 23,* 158–163.

Woolfolk, A. E. (1981). The eye of the beholder: Methodological considerations when observers assess nonverbal communication. *Journal of Nonverbal Behavior, 5,* 199–204.

Word, C. O., Zanna, M. P., & Cooper, J. (1974). The nonverbal mediation of self-fulfilling prophecies in interracial interaction. *Journal of Experimental Social Psychology, 10,* 109–120.

Wright, N. (Ed.) (1974). *Understanding human behavior: An illustrated guide to successful human relationships.* New York: BPC Publishing.

Wright, N., & Joiner, S. (Eds). (1974). *Understanding human behavior, An illustrated guide to successful human relationships,* Vol. 1. New York: Columbia House.

Yalom, I. D. (1970). *The theory and practice of group psychotherapy.* New York: Basic Books.

Zeiger, R. S. (1978). *Gaze aversion as an unobtrusive measure of homophobia.* (Doctoral dissertation, University of Maryland.)

Zeisset, R. M. (1968). Desensitization and relaxation in the modification of psychiatric patients' interview behavior. *Journal of Abnormal Psychology, 73,* 18–24.

Zimbardo, P. G. (1977). *Shyness: What it is, what to do about it.* Reading, MA: Addison-Wesley.

Zuckerman, M., Hall, J. A., DeFrank, R. S., & Rosenthal, R. (1976). Encoding and decoding of spontaneous and posed facial expressions. *Journal of Personality and Social Psychology, 34,* 966–977.

Zunin, L., & Zunin, N. (1972). *Contact: The first four minutes.* New York: Ballantine Books.

# Index